Dry Land Technology
Factors and Adoption

Dry Land Technology
Factors and Adoption

Chiranjiv Gandhi

RANDOM PUBLICATIONS
NEW DELHI - 110 002 (INDIA)

Dry Land Technology: Factors and Adoption

ISBN 978-93-51112-78-5

Published in 2014 in India by

RANDOM PUBLICATIONS

4376-A/4B, Gali Murari Lal, Ansari Road
New Delhi-110 002
Phone: +9111-43580356, 23289044
E-mail: randomexports@gmail.com; sales@randompublications.com; info@randompublications.com

Reprinted 2026

Type Setting by: Friends Media, Delhi-110089

Preface

Dry Land degradation and desertification now affect almost a billion people around the world. Tragically, the biological resources and productivity of millions of acres of land are lost to desertification each year because people remain unaware of strategies and techniques that could improve yields, reduce risk, and begin healing the world's deserts. As a result of the fast changing climatic conditions in the world, arable lands have been converted into semi arid, while formerly semi arid regions are now arid. However, the opposite is happening in Israel. The once arid land has slowly been converted into arable land, producing crops, fish and livestock that are exported to many other parts of the world today.

Dry Land farming has evolved as a set of techniques and management practices used by farmers to continually adapt to the presence or lack of moisture in a given crop cycle. In marginal regions, a farmer should be financially able to survive occasional crop failures, perhaps for several years in succession. Dry Land farming involves the constant assessing of the amount of moisture present or lacking for any given crop cycle and planning accordingly. Dryland farmers know that to financially succeed they have to be aggressive during the good years in order to offset the bad years. Research institutes have developed improved dry land technologies in order to increase yield of dry land crops. A widespread adoption of these improved dry land technologies by the farmers is vital for development of the economy of the country. Techniques include summer fallow rotation (in which one crop is grown on two seasons' precipitation, leaving standing stubble and crop residue to trap snow, and preventing runoff by terracing fields. Crop planning as per land use capability, Crop planning for aberrant weather situation in dry lands. Use of minimal irrigation: Moisture due to low rainfall and limited soil moisture due to soil depth are the situations normally experienced in dryland agriculture. Mid season droughts and soil moisture deficiency could be mitigated by applying protective irrigations to the crops alermn growth stages. The harvested water in farm ponds or from seam functioning wells can be utilized for the purpose. Besides putting various measures to increase the productivity levels of dry land crops, efforts would also be needed to increase the cropping intensity in dry land areas which was generally 100%, implying that a single crop was taken during the year. Cropping intensities of these areas could be increased by practice of inter cropping and multi cropping by way of more efficient utilization of resources. The cropping intensity would depend on the length of growing season which in turn depends on rainfall pattern and the soil moisture storage capacity of the soil.

The book discusses the ecology of desert plants, explores the causes of desertification and land abuse, and outlines the processes and procedures needed to evaluate, plan, implement, and monitor desert restoration projects. It sets forth economical and practical field-tested solutions for understanding site characteristics, selecting and growing plants, and ensuring that they survive with a minimal amount of water and care.

I thank all members of my team who have helped in the preparation of the book. My special thanks go to "Random Publications" who have published the book.

—Chiranjiv Gandhi

Contents

1

Introduction

Drylands

Drylands are defined by their scarcity of water. They are zones where precipitation is counterbalanced by evaporation from surfaces and transpiration by plants (evapotranspiration). UNEP defines drylands as tropical and temperate areas with an aridity index of less than 0.65. The drylands can be further classified into four sub-types: dry sub-humid lands, semi-arid lands, arid lands, and hyper-arid lands. Some authorities consider Hyper-arid lands as deserts (UNCCD) although a number of the world's deserts include both hyper arid and arid climate zones. The UNCCD excludes hyper-arid zones from its definition of drylands.

Drylands cover 41.3% of the earth's land surface, including 15% of Latin America, 66% of Africa, 40% of Asia and 24% of Europe. Worldwide there is a significantly greater proportion of drylands in developing countries (72%), and the proportion increases with aridity: almost 100% of all Hyper Arid lands are in the developing world. Nevertheless, the US, Australia and several countries in Southern Europe also contain significant dryland areas.

Drylands are complex, evolving structures whose characteristics and dynamic properties depend on many interrelated links between climate, soil, and vegetation.

Dryland Farming

By 2000 A.D., India will have to produce 240 million tonnes of food grains to feed her one billion population. This target cannot be realized from irrigated areas alone as we have irrigation potential for

178 million hectares only. Therefore, we will have to evolve an appropriate technology for dry land farming. On the other hand, we can say that second 'green revolution' in Indian agriculture can be had in rainfed/dryland agriculture. This is important to improve the standard of living of farmers residing in these areas as well.

India has about 108 million hectares of rainfed area which constitutes nearly 75% of the total 143 million hectares of arable land. In such areas crop production becomes relatively difficult as it mainly depends upon intensity and frequency of rainfall. The crop production, therefore, in such areas is called rainfed farming as there is no facility to give any irrigation, and even protective or life saving irrigation is not possible. These areas get an annual rainfall between 400 mm to 1000 mm which is unevenly distributed, highly uncertain and erratic. In certain areas the total annual rainfall does not exceed 500mm. The crop production, depending upon this rain, is technically called dry land farming and areas are known as dry lands.

India has about 47 million hectares of dry lands out of 108 million hectares of total rain fed area. Dry lands contribute 42% of the total food grain production of the country. These areas produce 75% of pulses and more than 90% of sorghum, millet, groundnut and pulses from arid and semi-arid regions.

Thus, dry lands and rainfed farming will continue to playa dominant role in agricultural production.

Dry lands, besides being water deficient, are characterized by high evaporation rates, exceptionally high day temperature during summer, low humidity and high run off and soil erosion. The soil of such areas are often found to be saline and low in fertility. As water is the most important factor of crop production, inadequacy and uncertainty of rainfall often cause partial or complete failure of the crops which leads to period of scarcities and famines. Thus the life of both human being and cattle in such areas becomes difficult and insecure.

Despite all these improvements in agriculture, we have yet not been able evolve an appropriate package of practices for our dry land areas. The income of farmers of dry land regions is still very low land areas. The income of farmers of dry land regions is still very low. To feed our one billion population that we will have by 2000 A.D., we will require food grains 10 the tune of 240 million tonnes approximately. For achieving this target we will have to harness every inch of our cultivable lands, esptX:ially dry lands, with utmost care.

Dry farming or dryland farming may be defined as: "a practice of growing profitable crops without irrigation in areas which receive an annual rainfall of 500 mm or even less". Efforts are being made to bring more area under irrigated agriculture and thereby to increase cropping intensity. But, even when we achieve our target of 113 million hectares of irrigated area by 2000 A.D., we would still have about 45% area under rainfed cultivation. We continue to stress on intensive agriculture on irrigated land but we can not afford to be complacent with our dry lands. Therefore, improved dry farming is necessary for equity and prosperity. As such we can not achieve stability in food production with unstabilized dry land agriculture. Therefore, we are required to ~ adopt improved technology especially developed for dry land agriculture.

Characteristics of Dryland Agriculture

Dry land areas may be characterized by the following features:

1. uncertain, ill-.distributed and limited annual rainfall;
2. occurrence of extensive climatic hazards like drought, flood etc;
3. undulating soil surface;
4. occurrence of extensive and large holdings;
5. practice of extensive agriculture i.e. prevalence of monocropping etc;
6. relatively large size of fields;
7. similarity in types of crops raised by almost all the farmers of a particular region;
8. very low crop yield;
9. poor market facility for the produce;
10. poor economy of the farmers; and
11. poor health of cattle as well as farmers.

Problems of Dry Farming in India

The major problem which the farmers have to face very often is to keep the crop plants alive and to get some economic returns from the crop production. But this single problem is influenced by several factors which are briefly described below.

Moisture Stress and Uncertain Rainfall

According to definition the dry farming areas receive an annual rainfall of 500 mm or even less. The rains are very erratic, uncertain and unevenly distributed. Therefore, the agriculture in these areas has become a sort of gamble with the nature and very often the crops have to face climatic hazards. The farmers also take up farming halfheartedly as they are not sure of being able to harvest the crops. Thus, water scarcity becomes a serious bottleneck in dry land agriculture.

Effective Storage of Rain Water

According to characteristics of dry farming, either there will be no rain at all or there will be torrential rain with very high intensity. Thus, in the former case the crops will have to suffer a severe drought and in the latter case they suffer either flood or water logging and they will be spoilt In case of very heavy downpour, the excess water gets lost as run-off which goes to the ponds and ditches etc. This water could be stored for providing life saving or protective irrigation to the crops grown in dry land areas. The loss of water takes place in several ways namely run-off, evaporation, uptake through weeds etc. The water could be stored for short period or long period and it can be preserved either in soil, pond or ditches based on situation and utilized for irrigation during dry periods.

Disposal or Dry Farming Products

In dry farming all the farmers grow similar crops which are drought resistant. These crops mature at the same time and the growers like to dispose off their products soon after the harvest. This results in a glut of products in the market and the situation is badly exploited by the grain traders and middlemen. Therefore, marketing becomes a serious problem in dry farming areas.

Selection or Limited Crops

Only drought resistant crops namely oilseeds, pulses and coarse grains like jowar, bajra, millets etc. can be grown in dryland areas. Thus, the farmers have to purchase other food grains and household commodities that unbalance their economic position.

Careful and Judicious Manurial Scheduling

In case of irrigated farming the farmers are at a liberty to apply [manures and fertilizers according to their availability and facility but in case of dry farming they have to be very careful in fertilizer application. Due to lack of available moisture, broadcasting or top dressing becomes wasteful and meaningless. These can be applied 'by only deep placement and foliar spray for an improved crop production.

Utilization of Preserved Moisture

Judicious and purposeful utilization of preserved moisture water depends upon soil type, plant type and other factors. The amount of available water to the plants depends upon the depth of plant roots, their proliferation and density. In case of limited moisture condition, the yield directly depends upon the rooting depth. The rooting depth can be desirably increased by mechanical manipulation of the soil. If the planting is very dense and all the plants have same kind of rooting then there will be a tough competition among roots for moisture and scarce moisture condition will result in the wilting of plants. Therefore, utilization of preserved moisture is an art in dry farming. The water collected in ponds or brooks may be used to give protective or life saving irrigation. The widely spaced crops can be intercropped with oilseeds or pulses for increasing the productivity of the land per unit area and per unit time. Therefore, the water. , collected during the rainy season need special technique and skill for its efficient utilization.

Quality or the Produce

The quality of the produce from dry farming areas is often found to be inferior as the grains are not fully developed or they are not filled properly; often mixed with other crop seeds owing to mixed .cropping system prevalent in these areas and the fodder become more fibrous. All these factors reduce the market value of produce and the farmers do not get the profit of their labour and Investment.

Work on Dry Farming in India

As it has already been stated, that crop production is highly risky in arid and semi-arid climates. In such conditions generally two types of agriculture is practised. One is crop production or arable farming and the other is mixed farming i.e. animal husbandry together with crop production and pasture management. But this type of agriculture holds true only in those countries where the population is limited and agricultural land is extensive such as Australia, South Africa and some states of the USA. In India, with high population and limited available land for agriculture, we have no other option than adopting arable farming.

The repeated and frequent crop failure in the past, resulting in short food grain supply, attracted the attention of our scientists and administration. Thus the scarcity of food grains in India were made the subject of enquiry in the year 1880 and the first Famine Commission was appointed in the same year. The commission after thorough study of the situation recommended the establishment of protective irrigation projects in South India and formation of department of agriculture in all the states. But nothing could be done illll923 when the first systematic and scientific approach to the dry farming problem in India was made.

Thus, as a beginning, Bombay research scheme on dry farming was started in 1934 at Solapur and Bijapur after the establishment of Indian Council of Agricultural Research (ICAR) a. New Delhi in 1929. After some time the work on dry farming wasi81so started in Punjab, Madras and Hyderabad. Dry farming work in U.P. started in Jhansi and Agra at dry farming centres established~ 1943-44 and 1948-50, respectively.

Since then the work has been in progress. Crop improvement brought out very promising strains during sixties which received our interest in finding out ways and means through which the crop production can be maximized in semi -arid and periodically dry areas of the country. Keeping this in view, the ICAR launched the All India Coordinated Research Project for Dry Land Agriculture in 1970 which was in active collaboration with government of Canada. The project started with multi disciplinary research units at 23 coordinating research centres located in various typical agro climatic regions of India with Hyderabad as headquarter.

The project started with the identification of the constraints responsible for lower yields in different regions : and then to develop a relevant location specific research programme to solve production constraints. Presently a joint team of senior scientists of India and Canada is working at Hyderabad.

To make the programmes more effective it is expanded to 16 agro-economic research centres. These centres have been established with an obvious objective of accelerating the conservation development and efficient, long term use of basic resources of soil and water for a self sustaining production. The main areas of investigations of these centres are given below:

- Identification of different crops and selection of high yielding varieties for different agro-climatic zones of the country.
- Developing cropping sequences and cropping systems suitable for dry farming.
- Determining the optimal crop population and planting pattern.
- Evaluating tillage implements and practices for water intake and storage in the soil profile, establishment of better crop stand and control of weeds.

- Designing and developing animal drawn implements for speedy and efficient cultural operations.
- Evaluating use of surface mulches, both organic and inorganic for short term moisture conservation.
- Determining ideal fertilizer doses and improving fertilizer use efficiency.
- Testing new planting rnaterials for introduction/substitution after they have proved their superiority over existing ones. - Harvesting and storing inevitable run off and recycling it as life saving or protective irrigation.
- Developing strategies for rneeting the challenges of an aberrant weather like skip or catch cropping etc.

Principal Dry Farming Zones in India

Almost all the states have some area under rainfed culture depending upon topography and irrigation facilities, but only the major dry farming areas are discussed here.

The Indo-Gangetic plains of North India

This zone is the youngest in the geological formation. This zone includes districts of Rajasthan, Punjab, Haryana, North- western M.P., and V.P. This zone is characterized by two major soil types namely light loam and heavy loam. The land is nearly leveled with a modest slope of 2 ft/mile length. The soils are very deep and situated at about 700 to 800 ft. above sea level. Because of heavy sand and silt fractions in the soil it has large pore spaces. The soils are rich in essential nutrients like nitrogen, phosphorus, potash, calcium etc. and, therefore, quite good for raising the crop excepting few with high water requirements. The cropping intensity, in this zone, stands around 120% and the major crops which are grown in this zone are millets, cereals, oil seeds and pulses. As far as rainfall pattern in this zone is concerned, it is observed that about 60% or more of the total rainfall is observed between the end of July to the end of August, and the rainfall in remaining months is quite poor. Thus, due to very high intensity of rainfall, floods are of frequent occurrence during the first week of September followed by a long spell of drought subsequently.

The Trapian Plateau of Peninsular India

This zone comprises the states of Maharashtra, Karnataka and Andhra Pradesh. The soil of this zone has been derived from the Deccan trap. The tract is undulating and consists of low ridges and valleys due to erosion which results in rapid run-off. About 40% of the land of this zone is not fit for cultivation. This tract is situated; at an elevation of 1400-2(XX) feet from sea level. The soil may be grouped into three types based on its depth as deep. medium deep and shallow soils. Leaching of lime has resulted in the formation of lime nodules or kanker on the surface soil. The soil is quite rich in total and available nitrogen. phosphorus and potash which favours production of crops if moisture is efficiently conserved.

In this zone. two high peaks of rain are observed because the area is affected by both south-west monsoon as well as northeast i monsoon. About 40-55% of total annual rainfall is obtained from south-west monsoon and the rest from north-east monsoon. Mostly the millets and some oil seeds like groundnut. are grown in this zone.

Plateau of Granite Formation

The soils of this zone are grouped as red soils and black cotton soils. Red soils are shallow while black cotton soils are very deep like clayey soils. The topography is of gentle undulations which favour run-off and soil erosion. The high pore space and high swelling of soil obstruct the permeability of rain water in to the lower layers of soil and its shrinkage results in hardening and clod formation on the surface which is unfavourable for plant growth. The red laterite and black cotton soils are deficient in nitrogen and phosphoric acids.

This zone also gets rain from two months namely south-west and north-east and the distribution pattern is more or less like peninsular group. Upland rice, millets, ragi, are the main crops of this zone. However. the yield of these crops is very low.

Steps for Raising Productivity in Dry Fanning

To boost the crop production under dry farming, we will have to efficiently manage our soil and water resources in the respective areas as dry land fanning gets more complex and intractable when droughts occur frequently. An efficient soil and water conservation system will playa vital role in boosting the crop yield in dry fanning. The different interdisciplinary approaches which are recommended for dry land fanning are categorized in to four major groups namely engineering, physiological, genetic and agronomic approaches.

Engineering Approaches

These approaches are aimed at soil and moisture conservation through regulation of run-off, collection of surplus rain water checking evaporation and seepage losses of water. and recycling of collected water as irrigation in times of critical need.

Contouring Across the Slope

Contouring is practiced on the lands with 3-5% slope. This system consists of constructing earthen bunds and the distance between the two bunds ranges from 30-50 m depending on the degree of slope. This is carried out with an object to provide a check to the flow of run-off water which then gets accumulated in the bunded area and is absorbed by the soil. Thus, contour bunding conserves moisture and prevents soil erosion.

Smoothening of Contour Inter-bund Areas

This is practiced only in those areas which have a slope of less than 1 per cent The smoothening may be achieved by running bullock drawn harrows or cultivators but small undulations are leveled during the process so that impounding of water may take place and maximum water absorption by the soil may be achieved.

Contour Border Strips Method

This method is suitable for areas having a slope of 3-4 per cent In this case parallel strips across the slope ranging from 10-15 m in width are laid down on contours and the soil surface is leveled by scrapping and placing the soil according to the need of the spot It is done to reduce the run-off and to conserve soil and water from the field. It is, however, an expensive method as it requires culling and filling up the soil from higher spots to lower ones.

Scooping or Land

In this practice, the land is generally scooped before the beginning of monsoon showers. By scooping, the soil is exposed for proper absorption and conservation of moisture. However, this is also a tedious as well as an expensive operation.

Opening or Ridges and Furrows

In this practice, the entire land is laid out into ridges and furrows across the slope. The ridges and furrows are opened before onset of monsoon so that the flow of water may be reduced and erosion may be controlled to the minimum. During rainy season, crops like maize, jowar, bajra, etc. may be grown in the furrows and legumes like soybean, arhar, urd, mung, cowpea, etc. may be grown on the ridges. After the monsoon is over the land is again leveled. This way the furrows are used to accumulate maximum water which will supply moisture for winter season crops.

Compartmental Bunding

Areas having a slope of 1% or less are suitable for compartmental bunding. It helps in accumulation of more water and a uniform spread of water in the entire area. Leveling is also done with nominal or no additional expenditure.

Bedding System

In this system, small furrows are opened and the soil from the furrows is uniformly spread in space left between the furrows. Thus, inter furrow spaces form the raised beds of about 4-5 metres width. This method helps in the conservation of soil, moisture and checking the excess run-off of water. The raised beds, in this practice, are used for growing such crops which need less water like legumes and oil seed crops, while the furrows are used for the crops which need more water.

Broad-based Bunding

This method is especially suitable for heavy black soil. Water is allowed to spread over a vast area by constructing a broad- based bund on a sloppy side. The water stays for a longer time because of high water holding capacity, lower leaching and seepage losses.

The stored water may be used for fish culture and also for [providing life saving irrigation grown in surrounding areas of catchment portion. These bunds are also called check dams and are I given a regulated drain or outlet for protecting the bunds from breaking.

Deep Summer Ploughing Followed by Surface Tilling

The field is ploughed deep by mould hard plough soon after harvesting rabi season crop with the objectives of (a) exposing the soil for perfect drying, (b) killing the disease pathogens, (c) destroying eggs of insect pests, and (d) controlling weeds by sun drying. The surface tilling during other seasons forms natural mulch and thereby reduces evaporation loss of water from soil.

Water Harvesting

Water harvesting is a technology of utilizing the collected and conserved water for the purpose of crop production. It includes tillage practices for an efficient use of moisture between and within the crop rows. Frequent stirring of the land by surface tilling provides mulch and prevents the evaporation loss of water from the soil.

Besides harvesting moisture from between and within the crop rows, run off losses are considerably reduced. The store water is used for providing life sowing irrigation to the crops grown in the surrounding areas. The water harvesting of this type can be done in areas situated near hill and on greatly undulated lands. In these cases, check dam tanks, and other reservoirs are constructed. The infiltration or percolation loss of water is prevented by spraying of asphalt compounds or by covering the bottom of the tanks or ponds through thin plastic sheets. The seepage loss may also be checked by providing a plastic lining. The evaporation loss of water is controlled by pouring some burnt crude oil over water surface. On a highly eroded soil or soils having very high slopes, terraces are made for providing a gentle flow of run off water so that the soil is not further eroded and some crop may be grown on these terraces.

Physiological Approaches

Hardly 1 % of the water absorbed by the plant roots is used for the growth and development of plants and remaining 99% is wasted through transpiration back to atmosphere. Thus one of the greatest causes of soil water wastage is loss of water through transpiration.

The extent of transpiration can be greatly influenced by using certain chemicals. These chemicals reduce transpiration, encourage root growth and protect the cytoplasmic proteins of the plants. These chemicals bring about more drought resistance in the plants. These compounds, according to their role, are classified as given below.

Anti-transpirants

Any chemical substance, which reduces rate of transpiration on its application to the plant surface, is called anti-transpirant. Any j substance which reduces the vapour pressure gradient in the stomatal cavity or increases stomatal resistance to water vapour diffusion, will act as anti-transpirant These substances have been used for arresting water loss from plant body with various degree of success. These are Phenylmercuric Acetate (PMA), Hydroxy Sulphonates (HS), Alkenyl Succinic Acid (ASA), Adol:-52 (a formulation of alcohol), and S-600 (a plastic transplanting spray).

Chemicals for Improved Cell Membrane Permeability of Water

Dry fanning areas are characterized by scarce rainfall and usually the roots have lipid layers which lower the absorption of water from the roots. Some chemicals like Alkenyl Succinic Acid (ASA) and Decenyl Succinic Acid (DSA), when applied, penetrate into the root and increase its water absorption power 8 times. Therefore, these chemicals are applied in the root zone for increased water absorption along with some chemicals to retard the transpiration from foliage.

Use of Plant Hormones and Growth Retardants

Some plant hormones like Indole acetic acid (IAA) and Abscissic acid (ABA) may be used for reducing the frequency and period of stomatal opening thereby minimizing the water loss from the plant body.

There are certain other chemicals known as growth retardants which either modify the plant structure or dwarf the plants by considerably reducing the total water requirement of the plants. The most important chemical of this group is cycocel or CCC (2- chlorocthyl trichloromethyl ammonium chloride). This chemical t\ also induces moisture stress tolerance in plants. Cycocel is presently used in cotton to encourage production of more fruiting branches (sympodial rather than monopodial or vegetative branches) and , Ii thereby even under drought condition it results in higher yield. In case of wheat, the use of Cycocel decreases the cell size and increases the density of the cytoplasm which ultimately results into a drought resistance into the plants.

Use of Chemicals

There are certain chemicals which are used for seed treatment to bring about drought resistance in plants right from seedling stage. Soaking of seeds in calcium chloride solution (0.25%) for 20 hours soaking of seeds with frequent shaking results in better germination and drought resistance in the plants.

Boron solution is also used for soaking seeds. Agrosan is a fungicide but also induces drought resistance in the plants when seeds are treated with this chemical.

Genetic Approaches

Because of scanty and unreliable rains the farmers of dry farming areas are still practicing crop husbandry on the basis of traditional approaches like low intensity cropping, little or no use of fertilizers or manures, raising low value crops. Crop varieties grown till the recent past were generally of long duration and slow growing. It, They were poor yielders too.

As nearly 70% of our total agricultural land is rainfed and 45% -of rainfed area is dryland, there is no way out but to evolve suitable varieties as well as appropriate technology for getting the most from our rainfed or dryland areas. As such, the concerted efforts of our plant breeders have resulted in the cultivation of several new plant types which possess all the characters needed for rainfed areas or dry lands. In terms of modem technology, such plant materials are called

"Ideo types". These are the suitable strains of dry land crops which are characterized by short growth duration, effective and extensive root system, drought tolerance, high yield potential having altered morphology of plants which are conducive to dry lands.

According to breeders, an ideal 'Ideo type" should have following qualities to give desired results in dry farming:

i. Early in growth duration and early vigour.

ii. Deeper root system with maximum branching at deeper zones.

iii. Dwarf plant types with lesser number of erect leaves.

iv. Moderate tillering; as profuse tillering causes competition. v. Good expression of ear heads even at higher planting density.

vi. Resistance to diseases.

vii. Bolder grains with moderate dormancy in them.

viii. Effective photosynthetic behaviour with greater sink capacity.

Agronomic Approaches

The major objective of dry farn1ingprogramme is to conserve the soil and moisture and to achieve maximum production from the dry farn1ing areas. In the past two decades, we have been able to' solve many hurdles in the aforesaid areas but there had been no break through as in case of irrigated crop production. Now we have the promising crop varieties and technology available with us about the maximum soil and water conservation.

The agronomic approaches can be dealt with under the following four heads based on land types.

Agronomic Approaches for Highly Undulating Lands

These lands are confined to the hills of the locations which have suffered serious soil erosion problems and have been divided into various gullies. The soils of these areas are more prone to further erosion if they are not properly managed. Therefore, in this category, the crop management practices are entirely different from other areas. The object of soil and crop management under such situation should be :

a. to stabilize the soil by forestry and pasture management with a regulated grazing or no grazing at all.

b. to level the land gradually through contour bunding, terracing, etc.

c. to practise strip cropping and pitcher farn1ing.

Agronomical Approaches for Marginal Lands

Generally, marginal lands are very poor in fertility. The crop management in these lands is carried out in lines of crop management for levelled lands or flat lands.

Agronomical Approaches for Diara Lands

Diara lands are located on either side of rivers or between two rivers and are often flooded by these rivers. These diara lands are formed due to flood and may have deposition of fine to coarse sands.

These areas often lack irrigation and need a careful crop management. Since the land of this areas is highly susceptible to floods, kharif cropping is practically impossible. But certain fodder crops can easily be grown soon after the onset of monsoon and harvested depending upon position of floods. The harvesting or cutting of crops is started from close to the river beds and as the water spreads the harvesting is also advanced. The life saving irrigations can be given by lifting water from the river or by drilling cavity wells or bam boo borings. However, in most of the cases these wells go out of order after a flood occurs. Therefore, the cavity wells or bamboo borings are made at a distant location from river stream.

Agronomical Approaches for Plain Lands

Plain lands form the main dry land tracts of the country. There has been major emphasis on finding out ways and means through which the total soil productivity could be increased. Following recommendations should be followed on plain lands for an improved crop productivity and an efficient soil and water conservation.

Tillage Requirements of the Crops

Tillage starts with the seed red preparation and ends with mulching and control of weeds. Deep ploughing during summer helps in destroying weeds and suppressing insect pests and diseases. It also helps in an efficient root penetration very deep into soil Placement of seed at 5 cm and fertilizers at 7.5 cm in the same furrow followed by soil compaction have resulted in better germination, plant vigour, extensive root development and higher crop yields.

Selection or Crops and Varieties

There are a number of improved varieties of different crops which are drought tolerant or resistant to water stress. The most commonly grown crops in dry lands are rice, maize, sorghum, pearl millet. finger millet, wheat, barley, pulses, oilseeds, etc. The improved varieties of these crops have already been described area-wise.

Sowing or Crops

Sowing of crops deals with several associated factors namely sowing time, method of sowing, depth of sowing etc. It is important in the sense that once the ideal plant population is achieved. the crop is bound to give yield.

Sowing time can markedly influence the production and productivity of dry land crops. Early sowing of kharif crops results in early crop maturity and thereby it facilitates early sowing of succeeding rabi crops. Early sowing of rabi crops helps in overcoming the moisture stress at later stages of plant growth, particularly at grain filling stage. Broadcasting of seeds should be avoided as it involves several losses and seed does not properly come in contact with moisture. Placing the seeds at about 5 cm depth through pora or seed drill is desirable. To get an ideal plant population it is necessary that about 25% higher than required seed rate should be applied. Care must be taken to reduce plant competition for moisture by removing excess plant population about 2-3 weeks after the sowing depending upon the crops.

Fertilizer Management

Use of fertilizers in dry lands is limited as compared to irrigated areas. Today we use on 'an average only 60kg/ha fertilizers in dry fanning areas as against 60 kg/ha national average. Reasons for application of fertilizers in dry fanning are as follows:

i. poor response because of faulty method of application;

ii. poor financial condition of fanners to purchase fertilizer;

ill. wrong concept of the fanners that fertilizers will burn the seedlings, and harm to the soil; and

iv. application of organic manure only, which can not meet the total nutrient requirement of the crop.

There are different Schools of thoughts about the application of fertilizers in the dry lands. Some people think that application of fertilizers in dry lands results in a better crop yield than that in irrigated areas because in the latter case the soil becomes very poor in residual fertility due to high intensity of cropping, whereas, the dry lands still preserve nutrients as they are not depleted badly due to low cropping intensity. Some people think that fertilizer application in dry lands aggravates the moisture problem, but truly speaking deep placement of fertilizers followed by foliar application results in extensive rooting up to deeper soil layers where the plants extract moisture to meet their demands.

Spreading of fertilizers at the surface or applying them at the seed layer is not so beneficial. Therefore, a careful and judicious fertilizer scheduling is a must frr achieving higher and better crop productivity in dry lands. Use of organic manures is always desirable but they alone are not enough as-they can not supply nutrients at the required speed and dose. Therefore, atleast half of the nutrient requirement of the crop must be met by readily available source of nutrient i.e. fertilizers. This also improves water use efficiency.

Following points should be considered to improve the fertility status of dry lands.

- Green manuring should be practised in kharif season.
- Inclusion of a legume crop in rotation adds about 20-25 kg of nitrogen per hectare, reduces cost of fertilization and increases crop productivity.
- Inoculation of legumes through Rhizobium culture and nonlegumes with Azotobactor adds about 30-35 kg of nitrogen per hectare.
- Foliar application of urea having lowest biuret content along with micro-nutrient fertilizers like zinc sulphate will increase grain yields in dry farming.
- Growing forage legumes in rotation with pearl millet in kharif season will provide a better condition for rabi crop.

Therefore, proper fertilizer scheduling will help the farmers to induce drought tolerance/resistance in the crops and provide - higher yields.

Cropping Systems

Cropping system refers to an arrangement in which various crops are grown together in the same field. The cropping systems followed in dry lands differ from those followed under normal conditions. Only those crops can be grown under dry land conditions which require less water to complete their life cycle or which can stand or yield under drought conditions. This can include both drought resistant and drought tolerant plants. In addition, plants can be grown only where some water is available to sustain the growth of plants. Following are a few intercropping systems for dryland areas:

Moong + Bajra

Guar + Bajra

Til + Guar/moth/mung

Mixed cropping is also followed to minimize the effect of unpredictability of rain. Mixed cropping may have low yield potential but it works as a buffer against failure under possible unfavourable conditions. Mixed cropping may be defined as sowing of two or more crops simultaneously on the same piece of land in separate rows. Examples: Guar + Arhar + Moong, Bajra + Arhar + Moong and Maize + Urd etc.

Cropping Pattern

Cropping pattern is defined as sequence of growing crops in a particular field at a particular period. The most common cropping pattern for dry land farming are discussed below:

For North Indian Conditions

- Sorghum -Safflower/mustard
- Sorghum –Mung/urd/cowpea -Gram/wheat -Gram
- Rice -Gram (for low lying areas)
- Bajra -Gram + Linseed
- Bajra + Urd/Mung/soyabean -Wheat/barley +
- Gram/m ustard
- Maize-Gram/safflower

For Central Indian Condition

- Green gram -Rabi sorghum
- Green gram -Safflower

For Bhubaneswar Region

- Rice -Horse gram
- Ragi -Red gram
- Groundnut -Niger
- Maize-Niger

Weed Control

Presence of weeds in the crop field, especially in case of dry lands, cause a severe crop weed competition for water, nutrients and light The reduction in yield due to weeds varies from 30-75% depending upon the crop and nature and extent of weed infestation. Weeds may be controlled by hand weeding, intercultural operations and herbicidal application or by adopting an integrated approach.

Plant Protection Measures

In light textured soils of arid and semi arid regions termites and white grubs cause extensive damage to emerging seedlings and also to grown up plants. Use of BHC 10% dust@ 25-30 kg or Aldrin 5% dust@ 10-15 k/ha, in the soil and incorporating it well into the soil at the time of the last ploughing will control termites. The white grubs may be controlled by drilling of Thimet 200 granules @ 15 k/ha along with seeds. Aphids in mustard are very destructive, therefore, they are controlled by spraying 0.2% Metasystox or Dimecron. Similarly, the pod borers in pulses are controlled by spraying 0.05% Endosulfon.

The viral diseases of pulses should be controlled by seed treatment and spray of some fungicides to kill the insect vectors.

Agro-forestry and Dryland Agriculture

Agroforestry is the cultivation of trees in association with crops. It has assumed a great importance in order to solve a number of problems of dryland farmers. The twin objectives of Agroforestry are to increase the farmers income and sustain the ecological! Environmental balance. Agroforestry, as it provides fuel wood to the fanners, enables them to save the animal dung and use it as manure. The following points should be considered while choosing trees for cultivation in association with crops:

- The tree should grow fast and stand repeated prunings.
- Its roots should explore soil layers not tapped by agricultural crops.
- The crown should not be too large to prevent sun light from reaching the crop.
- The peak period of flush and leaf faII should not be detrimental to the crop grown in association.
- As far as possible, the tree species should be a legume, the root nodules of which accommodate nitrogen manufacturing bacteria.
- The branching should be sparse and light.
- The tree should meet the socio-economic and ecological needs of the region.

Some agro-forestry combinations are given for the main dry fanning areas of the country. Trees suitable for growing in association with crops in different dry land regions are given below:

For North-western Arid and Semi arid Zones

Prosopis cineraria (khejri), Zizyphus mauriliana (jujube), Acacia IOrlilis(israely babool),Leucaena leucocephala (su-babool), Acacia auriculiformis.

For Central Arid Vindhyan Zone

Acacia nilolica var. cuprissiformis, Punica granalwn (anar) .Kinnow (citrus), Psydiwn guajava (guava).

South Eastern Regions 1

Acacia albida, Acacia ferruginea, Prosopis cineraria, Leucaena leucocephala.

Himalayan Regions

Alnus neplensis (utis), Sehima wallichianum, Citrus species, Sesbania sesban, Prosopsis juliflora

Recommendations for Dry Farming Areas

The research programmes of all India coordinated research projects for dry land agriculture have concluded into certain recommendations to the farmers of dry land areas which are described below.

1. Bunding across the slope and leveling the land should be done before onset of monsoon.
2. Deep summer ploughing should be followed by surface tillage during monsoon months and also rest of the year.
3. Application of organic manures like FYM compost. etc. @ 15-20 tonnes/ha or green manuring should be done. These manures should be applied about 20-25 days before sowing and should be well mixed in the soil.
4. Fertilizers should be basal placed at a depth of7.5 to 10cm in the soil and the seeds should be sown in the same furrows about 3 cm. above the fertilizers. This is important especially during winter season. The nitrogen (20-50% of total) should be top dressed by side or band placement method at about 10-15 cm apart. The crop rows should be done soon after the rains but if there is not sufficient moisture in the soil, the nitrogen should betrayed over the foliage with urea solution containing 3-5% nitrogen. Zinc and sulphur should be applied as basal if needed.

5. Soil application of BHC (10%) dust @ 25-30 kg/ha for termites and Thimet 20 G @ 15 kg/ha for white grub should be done. These chemicals must be mixed with soil properly while ploughing or at the time of sowing.
6. Selection of suitable crops and their varieties should be done I according to their suitability to a particular region/micro climate.
7. Seeds must be treated with a suitable fungicide and that of legume with Rhizobium culture before sowing. Soaking seeds in plain water for rabi sowing helps in getting higher gem1ination, better seedling vigour and an early maturity within weeks time.
8. Proper crop rotation should be followed which should preferably have at least one legume every year.
9. For better seed soil moisture contact through soil compaction should be done by running a plank or roller especially for rabi crop.
10. At the event of total crop failure during kharif season a suitable catch crop like urd (T-9) or toria etc should be sown.
11. Intercropping of oil seeds and pulses should be done with jowar, bajra and maize crops for the purpose of making best use of soil and inter row moisture harvesting.
12. Line sowing by drilling the seed at a depth of7.5 to 10cm or even more depending upon the situation should be practiced because it helps in better seed germination. This also helps in stabilizing the required plant population and thereby in getting better yield.
13. Proper weed management practices should be followed by adopting integrated weed control measures.
14. Mulching should be done by providing frequent interculture and pulverizing the soil. If intercultural operations are not possible then use of artificial mulches like covering the surface with tree leaves, uprooted weeds, sugarcane leaves, saw dust or polythene sheets are used to check the evaporation of water from the soil.
15. Water harvesting between the rows should be done by growing some pulse crops and run off water should be collected in some nearby located ponds and used as life saving irrigation.

16. An efficient plant protection measure should be adopted to protect the crop from insect pests and disease damage.
17. The crop should be harvested at proper physiological maturity so that the following or succeeding crop may be sown slightly earlier than the scheduled time and best use of rain water or residual moisture may be made for crop production.
18. Crops like cotton, chilies, etc. should be sprayed with CCC or cycocel and groundnut should be sprayed with planofix for modified growth, higher drought resistance and better yield.

Conclusion

Even after utilizing all the available water resources, about 50% of our culturable area will still depend on rains. Therefore, our agricultural scientists, policy formulators and farmers should appropriately realize the magnitude of role that. rainfed agriculture or dryland farming can play. They should thoroughly examine the problems of dry land agriculture from different view points and evolve appropriate technologies, crop varieties, etc. for these areas to better the economic position of the farmers. Dry farming areas, therefore, need a much closer attention.

Farmers should utilize well in time whatever improved technology and varieties suitable for dry farming are available. They should be extra careful about the utilization of available rain water, selection of crops and protection of crops from different harmful physiological or biological agencies. And only then they can make dry farming as profitable as irrigated farming itself.

2

Fertilizer Use in Dryland Farming

To meet the increasing food grain requirement of the country, dry lands will have to playa crucial role. A large chunk of country's poor population inhabits the dryland areas. An attempt to improve the productivity of dryland areas intend to serve two purposes: improving the economic status of poor masses deriving their livelihood from dryland areas and contributing to the total food. grain production of the country. Fertilizers have a key role to play in improving productivity of dryland areas. To ensure the benefit and minimize the risk, a proper understanding of certain theoretical and practical aspects of fertilizer application in dryland areas is necessary.

Nearly 70 % of the arable land in our country is rainfed. This large chunk of area contributes only 42 % of the total food grain production. The scarcity of water in these areas greatly inhibits adoption of modern package of practices in farming including the use of high yielding varieties, pesticides and fertilizers. Consumption of fertilizers per hectare in dryland regions is nearly 3 times less than that in irrigated areas. More than 80 % of the total fertilizers used in the country is dumped in only 30% irrigated area. Researches have, however, proved that the productivity of dry lands can be improved significantly by adopting appropriate method of fertilizer application.

Dry lands have not only a thirst for water but also a hunger for nutrients. Nitrogen deficiency is a widespread problem of dryland areas. Low amount of moisture in the soil and relatively higher temperature encourage the loss of nitrogen from the soil. Soil nitrogen is lost to the atmosphere in gaseous form. Deficiency of phosphorus, potassium, sulphur and zinc is also frequently observed in dryland areas. These nutrients are essential for the growth and development of plants.

Nutrient Availability

Nutrient availability refers to the amount of nutrients present in the soil which plants can easily absorb by their roots. A nutrient is available, only when it is soluble in water. In the regions of adequate rainfall/irrigation, main thrust is to make the nutrients more and more soluble in the water, so that the plant roots may absorb them. But in dryland areas problem is just contrast. These soils do not have enough moisture to dissolve plant nutrient, even if the nutrient is present in the soil in water soluble form. Therefore, nutrients in the soil do not find their way into the plant roots. Application of fertilizers is, therefore, highly restricted in dryland areas.

Recommendation of Nutrients

Out of three primary fertilizer nutrients namely I nitrogen, phosphorus and potassium, nitrogen has the highest bearing on the crop yielding in dry land farming. This is because of the fact that dryland soils are highly deficient in nitrogen. Response of nitrogenous fertilizers is always greater than phosphatic and potassic fertilizers. On the other hand more amount of nitrogen can be made available than phosphorus and potash at a definite level of soil moisture. Nitrogen is more mobile in the soil than phosphorus and potash. A definite amount of moisture can transport more amount of nitrogen than phosphorus and potash. Yet, the amount of nitrogen recommended for crops in dryland condition is always less than in irrigated conditions. Generally less amount of fertilizers is recommended for cereal crops in dryland condition than in irrigated conditions. But for leguminous crops there is not much difference. Recommendation of doses for main crops grown in dryland areas.

Table: *Nutrient requirements of various crops in various regions.*

Sl No.	*Region*	*Crop*	*Quantity of nutrient kg/ha*		
			N	P	K
1	2	3	6	5	6
	1. Kharif black soils of Jhansi(U.P.)	Sorghum	0	40	0
		Red-gram	0	40	0
		Sesamum	0	30	0
		Guar	0	60	0
		Groundnut	0	25	0
		Wheat	0	30	0
		Barley	0	30	0

	Linseed	0	20	0
	Mustard	0	20	0
	Safflower	0	20	0
2. Kharif Black soils of Rajkot (Gujarat)	Groundnut	0	25	0
	Sorghum	0	30	0
	Pearl miller	0	20	0
	Cotton	0	25	0
	Sesamum	0	25	0
	Castor	0	40	0
	Green-gram	0	40	0
	Pigeon pea	0	40	0
3. Kharif and rabi Black soils of Sholapur (Maharashtra)	Bajra	50	25	0
	Rabi Sorghum	50	0	0
	Safflower	50	25	0
	Bengal gram	15	25	0
4. Indore(Madhya Pradesh)	Sorghum	50	25	0
	Maize	50	25	0
	Wheat	30	30	0
	Safflower	40	40	0
	Bengal gram	20	40	0
5. Udaipur(Rajasthan)	Maize	50	30	0
	Sorghum	50	30	0
	Bengal gram	15	30	0
	Wheat	30	15	0
	Safflower	30	15	0
	Mustard	30	15	0
6. Rabi black soils Bellary Karnataka	Rabi Sorghum	30	30	0
	Safflower	20	30	0
	Bengal gram	15	30	0
	Field bean	15	30	0
7. Kovilpatti(Tamilnadu)	Sorghum	40	20	0
	Pearl millet	40	20	0
	Pulses	20	40	0
8. Alluvial soils Agra(U.P)	Bajra	60	30	0
Black&Red gram	10	40	0	
	Guar	10	20	0
	Mustard	60	40	0
	Safflower	60	40	0
	Bengal gram	10	25	0
	Barley	60	0	0
	Wheat	40	0	0

	Linseed	40	20	0
9. Semi-arid Red soils(A.P)	Groundnut	20	40	40
	Castor	40	40	40
	Red gram	20	40	20
	Bajra	40	40	40
10. Hyderabad(Andhra Pradesh)	Sorghum	40	30	0
	Castor	50	30	0
	Red gram	10	30	0
	Ragi&Bajra	40	30	0
11. Bangalore (Karnataka)	Ragi	50	50	25
	Maize	75	50	25
	Groundnut	25	50	25
	Pulses	25	50	25
12. Sub humid red soils Bhubaneshwar (Orissa)	Upland rice and medium rice	75-90	60	40
	Barley	30	20	20
13. Ranchi (Bihar)	Rice (HYV)	60	30	0
	Rice (local)	30	20	0
	Wheat	30	20	0
	Safflower	20	0	0
	Bengal gram	0	20	6

Out of the three primary nutrients, nitrogen and phosphorus are applied in significant amount. In scanty rainfall areas, potassium seldom becomes a limiting factor jn crop yield. Its use is, therefore, avoided in most crops.

Improvement of Fertilizer Use Efficiency

The objective behind the use of fertilizer in dry land condition should be to make maximum amount of fertilizer nutrient available to the plant within the existing low levels of soil moisture. Certain points, discussed below, must be kept in mind in order to achieve this objective.

Use of Organic Matter

Organic matter constitutes a vital component of the soil. It is made up of plant residues, animal residues and dead and decayed soil organisms. To maintain a fair level of organic matter in the soil, use of farmyard manure, green manure, compost, and other organic manures is essential. The size of well decayed organic matter particles is more or less equal to the finest particles of the soil called clay. But organic matter particles play more important role than clay in retention of soil moisture. A unit volume of organic matter particles can retain 10-100 times greater amount of moisture than the same volume of clay. Use of materials supplementing to organic matter reserves of the soil is therefore, imperative in dry land soils. Besides augmenting moisture retention in the soil, organic matter helps in the development of granular soil structure which provides good aeration to the soil. Roots find congenial atmosphere for proliferation it and can extract greater amount of moisture and nutrients from the soil.

Green manuring in assured rainfall areas is very useful. Legume crops enrich soil fertility by fixing nitrogen and supplying organic matter to the soil. Growing of green gram, cowpea, black gram and soyabean with early monsoon showers for grain and by incorporating residues in the soil for August sown sorghum is a useful practice. On roadsides and field boundaries which are lying unused, leguminous shrubs and trees can be grown and the leaves and soft twigs can be cut, collected and transported to the main field. About 10 to 30 kg of fertilizer nitrogen per hectare can be saved if green manuring is done for the main crop.

Placement of Fertilizers

Placing the fertilizers in the root zone of the soil is a more efficient techniques than broadcasting of fertilizers. In the placement application, fertilizer is placed in furrows behind the desi plough at appropriate depth and at safe distance from the seeds or seedlings. By this method, contact between fertilizer and soil particles is reduced and unavailability of fertilizer nutrients due to their fixation with the soil particles can be avoided to a large extent. Fertilizer nutrients can manage to reach plant roots even if moisture in the soil is scanty.

Technique involves opening furrows towards one or both the sides of the row of crop plants, placing the fertilizer in these furrows and covering them. Furrows can be opened with the help of bullock drawn plough. Some bullock drawn machines called 'seed cum fertilizer drill' are also used for this purpose in some parts of the country. In orchard trees, fertilizer is placed in a ring which is dug around the tree at a safe distance from the tree trunk.

Spray Application

Spray application of urea and micronutrient fertilizer to supply nitrogen and micronutrients respectively, to the plants has been found to be very useful in dryland conditions. When moisture in the soil is not sufficient to transport fertilizer nutrient from the soil to the plant roots, use of, spray solution is desirable. This is an effective method of application of nitrogen through urea. Spray grade urea is marketed in our country. A 2-3% solution of urea in water can be sprayed on the crop. Two to three sprays of this kind can be given to the standing crop. If a farmer fails to apply fertilizer to his crop in time due to some circumstances such as lack of timely rainfall, unavailability of fertilizer in time and inability to farmer to buy fertilizer in time etc., application of 2-3 % urea, solution proves very beneficial.

Balanced use of Fertilizers

The singular application of either nitrogen or phosphorus or potash is not as useful as the application of their combinations. Use of phosphatic fertilizer along with nitrogenous fertilizer is found specially beneficial in dryland conditions. Phosphorus helps in elongation of root network of the plant, thereby enabling the plant roots to explore greater volume of ambient soil for nutrient and moisture. Elongated roots absorb greater amount of nutrients from the soil like nitrogen, potash, calcium etc. On the other hand, nitrogen along with other nutrients helps in development of sound foliage. The ultimate impact of the combined use of nutrients is the higher yield of crop.

Use of Phosphorus Mixed with Farmyard Manure (FYM)

Farmyard manure prepared by this method is called enriched FYM. About 750 kg of FYM is sufficient for one hectare. The whole amount of phosphatic fertilizer, calculated on the basis of recommended dose of phosphorus, for a planned crop is mixed with farmyard manure before its application to the field. Fertilizer and FYM are spread in alternate layers over one another and left in the manure pit for ripening for about a month. Pit should be covered with soil to make it airtight. Enrichment of FYM by phosphorus also increases nitrogen content of FYM. Phosphorus of FYM does not allow gaseous loss of nitrogen from FYM as it absorbs nitrogen and withholds it within the manure. Use of enriched FYM has been found very useful in dryland soils.

Use of Micronutrients and Amendments

Micronutrients are essential for the plants though the plants require them in very minute amounts. Sometime, plant growth becomes impossible due to the absence of micronutrients in the soil. Crop yield is drastically reduced due to the deficiency of micronutrients and use of fertilizers does not respond significantly. In such a situation, a representative soil sample should be taken to a local soil testing laboratory and tested there for nutrient status of the soil. Imbalance of nutrients is very likely when soil reaction is not neutral. Soil testing report is supposed to give information about soil reaction, accordingly, appropriate amount of lime in acidic soils and gypsum in alkali soils should be applied. Nutrient re- commendations based on soil test are always more desirable than blanket recommendations.

Use of bio-fertilizers

Use of bio-fertilizers is very beneficial in dry land situations. Treating the leguminous crop seeds with Rhizobium bio-fertilizers increases capacity of the crop to fix atmospheric nitrogen and dependence of plant on soil nitrogen is greatly reduced. Roots and other residues of legume crops left in the soil after harvesting, enrich the soil with nitrogen and subsequent crops get good amount of nitrogen from this soil. Mycorrhiza which is a promising biofertilizers may prove very beneficial for supporting phosphorus to crops. This biofertilizer consists of an organism called fungi. The body of micorrhizal fungi. is like a network of fine threads called hyphae. Hyphae get wrapped over plant roots and extend into the soil solution. Hyphae absorb plant nutrient from the soil, transport them to vascular bundles of the plant roots and leave them there for plant use. Application of other bulky bioertilizers. like blue green algae and azolla is also helpful in retention of moisture and increasing recovery of fertilizer used.

Alteration in Time of Fertilizer Application

Crops need fertilizer at a particular stage of growth. But a slight modification in time of fertilizer application based on moisture availability in soil, is desirable. If rainfall takes place shortly before or after the usual time of fertilizer application can be made to suit the nutrient availability. Fertilizer should be applied when there is adequate moisture in the soil. This results in greater utilization of nutrients by plants. In those dry1and fields where irrigation facility is limited, calibration in timings of fertilizer application can be made accordingly.

Water Conservation Practices

It is obvious from the term dry1and farming that, water deficiency is a perpetual problem and the biggest constraint to adoption of fertilizer use technology. Water conservation practices tend to encourage fertilizer application in dryland areas. All the measures should be adopted to arrest every drop of rainwater on the field itself. This will ensure greater utilization of native plant nutrients and consequently greater yield of crops.

In those dry land areas where soil and water conservation practices are adopted, fertilizer consumption is greater than that in the areas with no water conservation practices. Beneficial effect of fertilizer application coupled with moisture conservation practice. Dose of fertilizer recommended for wheat in dryland condition was applied; moisture level was measured as thickness of water layer in millimeter, in the top 90 cm thick vertical layer of soil.

Table: *Effect of moisture conservation practices on wheat yield*

Sl.No	*Treatment*	*Moisture level in soil profile (mm/90cm)*	*Yield as average of two years (kg/ha)*
1	Farmers practice	147	1500
2	Minor leveling& bundling	168	1780

Graded bunding with water ways should be preferred ~ for all black and heavy textured soils like clay, clay loam and silty clay. Graded bunding is also suitable for all areas receiving more than 800 mm of rainfall and soils having impervious layers within a depth of 22.5 cm irrespective of soil type. Contour bunding is recommended for areas receiving less than 800 mm rainfall and having light to medium texture soils with good permeability. For the slopes, terracing is recommended. Whereas in the cultivated land deep ploughing is an easy method of water conservation.

Other Farming Practices

Other farming practices should be based on two basic necessities of dryland farming namely moisture and nutrient supply. In kharif season nitrogen should be given in two or three split doses in order to avoid leaching loss of nitrogen. However, for rabi season crops, nitrogen can be placed along with phosphorus and potash before sowing. Proper sequence of crops (rotation) help in balanced use of nutrients in the soil. Crop rotation must include at least one leguminous crop.

Certified quality seeds should be preferred to traditional ones. It is often observed that the high yielding varieties of sorghum, pearl millet, castor, and beans, adapted to that particular region, have deeper and more extensive root systems than the local varieties. Hence, they are able to withstand moisture stress and absorb greater amount of nutrient from the soil.

Some farming practices specific to dryland conditions like manipulation in sowing time, following cropping pattern and cropping intensity recommended for that particular area, proper insect pest and weed control etc. must accompany the use of fertilizers.

Mulching

The practice of mulching involves placing materials such as straw, plant residues leaves, loose soil or plastic film on the soil surface to reduce loss of moisture by evaporation from the soil. Mulching reduces soil erosion and also protects the plant roots against wide fluctuation in soil temperature.

Mulching of rainfed crops helps retain soil moisture and increase fertilizer use efficiency. Yield of wheat often increases if mulching is done in its preceding crop of maize. In wheat, following mulched maize, the yields with 0 and 40 kg N/ha were comparable to those with 40 and 80 kg N/ha without mulching. Thus, it saved 40 kg N/ha to the succeeding crops of wheat.

Conclusion

To meet the food grain requirement of 200 million additional population by 2000 A.D., we have to produce 70 million tonnes of food grains over and above what we produce today (target of 2000 A.D. is 240 million tonnes).

More than 80 % of the additional requirement of food has to be met from the dryland area in nineties. Proper application of the appropriate technology, of proven (benefit, must receive attention. The package of practices of dryland farming evolved by the various research institutions located in various parts of India have made the fertilizer use technology free from risk, cost effective and ecologically sound. Farmers in dryland areas are very receptive to fertilizers.

Yet, a large number of them stay away from fertilizers as the use of fertilizers involves a risk because 1 of uncertainty of monsoon in India and a very short 1 span (only 2-3 months) of rainfall. The mean length of growing season varies from 18 weeks on shallow red soils, 22 weeks on deep red soils to 22 weeks on deep black soils.

The calendar of operations should be so framed as to take the fullest advantage of growing season.

Investigations have shown that in assured rainfall zones of dry land regions, investing one rupee in fertilizer gives return of 2 to 15 rupees in monsoon season.

In post monsoon season the return has been worked out to be 2 to 12 rupees for each rupee invested in fertilizer, in assured rainfall zone.

To realize this much profit, fertilizer application must be accompanied by other improved practices of dry land farming.

Improved Dryland Technologies and Weed Management

Intercropping: Mixed cropping is traditionally adopted in rained areas for risk distribution during adverse weather conditions there by achieving yield stability.

Now the mixed cropping has changed to intercropping due to substantial research efforts. The intercropping has proved substantial yield advantage over sole crop. It also gives greater stability of yields over different seasons.

The intercropping is more popular among the small farmers to meet their need for different crops. The intercropping is more advantageous during Kharif than rabi season.

Points to be considered while selecting crops for intercropping:

i) Both the crops should have differential growth habit

ii) They should have different crop growth period.

iii) There should be more competition tree period between these two crops.

iv) They should have differential requirements interspect at nutrients and soil moisture.

v) They should have different feeding zones.

vi) Mostly there should be a combination of cereal and legume.

Eg .Bajra + Tur - 2:1 proportion

Cotton + green gram - 2: 1 proportion

Sorghum + green gram 2: 1 proportion

Sunflower + Tur - 2: 1 proportion

Sorghum + Tur 2: 1 proportion

Fertilizer Use in Dryland

Fertilizer use is an reertant factor for increasing the yield of dryland crops. The soils of drylancare low in N low to medium in P and rich in available K.

The research breeds out in fertilizer use of various crops have suggested ad definite resense to fertilize, application. The fertilizer application not only increase are vield of dry land crops but also enhance the growth and maturity of the nops.

Based on the results of various trials conducted at Solapur the nacwing economic fertilizer doses are recommended for dryland crops.

Crops	*Recommended fertilize does (Kg/ha)*		
	N	*P205*	*K2S*
A) Scarcity zone (Solapur)			
Kharif			
1. Bajra	50	25	-
2. Tur			
3. Gr. Nut	12.5	25	-
4. Horse gram			
5. Kidney bean			
6. Sunflower	50	25	-
7. Castor	25	12.5	
Rabi			
1. Sorghum	50	-	-
2. Safflower	50	25	-
3. Gram	12.5	25	-
B) Assured rainfall zone (Akola)			
1. Jowar	80	40	-
2. Rala	60	30	-
3. Cotton	80	40	-
4. Tur, Mung, Urid, Gr. Nut	20	40	-

The time and method of application of fertilizer are also imcutant because of moisture limitation of dryland. The fertilizer should be applied at the time of sowing by placement method. For this purpose, Solapur centre has developed two bowl ferti seed drill which now become more sepular among the farmers. The response top application to deep black soils is limited the to high p fixation due to calcium content. For Rabi crops advance placement of p by 10 - 20 days before actual sowing is usual practice. The soils of dryland are medium to rich in available K and hence no sponse to K application.

Under light textured soils boron application

@ 5 kg Borax x / ha at in ernate years to groundnut is recommended for dry lands of Maharashtra.

Under intercropping systems, there is no need to apply additional utilizes to intercrop of legume under cereal + legume combination.

Under sequence cropping under dry lands, when cereal is grown after gumes like green gram, black gram, the N application may be reduced by 5 to 30 kg N / ha to cereal or oil seed crops like sorghum sunflower of safflower.

Weed Management: It is estimated that weed cause about 37 to 79% pass to the crop production in dryland agriculture where moisture is the most limiting factor. The research findings on weed - crop competition in Dryland indicated that in crops with 80 to 120 days duration first 20 to 30 lays period is more sensitive to competition from weeds. Checking weed growth during this period helps in to minimize the loss in crop production. The results with perarlmillet at Solapur confirm the above observations.

Weed control through chemicals: Chemicals are found to be as effective as mechanical measures in controlling the weeds in dryland agriculture. Use of Atrazine @ 0.5 to 1.0 kg a.i. / ha as pre-emergence for cereal crop is effective. The weeds like striga can be controlled effectively by 2, 4 - D from mechanical measures.

Integrated weed control: A combination of mechanical and chemical weed control methods are most effective. Studies carried out at Solapur on methods are most effective. Studies carried out at Solapur on Kharif groundnut showed that combination of pre - emergence spray of Propanil @ 3 kg/ha + weeding at 30 days after sowing increased yield of groundnut by 100 per cent. Use of Basaline ! 1 kg/ha + one hand weeding at 30 DAS has also found effective at other locations.

Off Season Tillage for Weed Control: Off season tillage would control the weeds reaching maturity after the harvest of first crop and checking multiplication. It also helps in timely sowing of crop with first shower of rain.

Rainfed Agriculture

The term Rainfed agriculture is used to describe farming practises that rely on rainfall for water. It provides much of the food consumed by poor communities in developing countries. For example, rainfed agriculture accounts for more than 95% of farmed land in sub-Saharan Africa, 90% in Latin America, 75% in the Near East and North Africa; 65% in East Asia and 60% in South Asia.

Levels of productivity, particularly in parts of sub-Saharan Africa and South Asia, are low due to degraded soils, high levels of evaporation, droughts, floods and a general lack of effective water management. A major study into water use by agriculture, known as the Comprehensive Assessment of Water Management in Agriculture, coordinated by the International Water Management Institute, noted a close correlation between hunger, poverty and water. However, it concluded that there was much opportunity to raise productivity from rainfed farming.

The authors considered that managing rainwater and soil moisture more effectively, and using supplemental and small-scale irrigation, held the key to helping the greatest number of poor people. It called for a new era of water investments and policies for upgrading rainfed agriculture that would go beyond controlling field-level soil and water to bring new freshwater sources through better local management of rainfall and runoff.

The importance of rainfed agriculture varies regionally but produces most food for poor communities in developing countries. In subSaharan Africa more than 95% of the farmed land is rainfed, while the corresponding ûgure for Latin America is almost 90%, for South Asia about 60%, for East Asia 65% and for the Near East and North Africa 75% (FAOSTAT, 2005). Most countries in the world depend primarily on rainfed agriculture for their grain food. Despite large strides made in improving productivity and environmental conditions in many developing countries, a great number of poor families in Africa and Asia still face poverty, hunger, food insecurity and malnutrition where rainfed agriculture is the main agricultural activity. These problems are exacerbated by adverse biophysical growing conditions and the poor socioeconomic infrastructure in many areas in the semi-arid tropics (SAT). The SAT is the home to 38% of the developing countries' poor, 75% of whom live in rural areas. Over 45% of the world's hungry and more than 70% of its malnourished children live in the SAT

There is a correlation between poverty, hunger and water stress (Falkenmark, 1986). The UN Millennium Development Project has identiûed the 'hot spot' countries in the world suffering from the largest prevalence of malnourishment. These countries coincide closely with those located in the semi-arid and dry subhumid hydroclimates in the world, i.e. savannahs and steppe ecosystems, where rainfed agriculture is the dominating source of food and where water constitutes a key limiting factor to crop growth (SEI, 2005). Of the 850 million undernourished people in the world, essentially all live in poor, developing countries, which predominantly are located in tropical regions (UNSTAT, 2005).

Since the late 1960s, agricultural land use has expanded by 20–25%, which has contributed to approximately 30% of the overall grain production growth during the period (FAO, 2002; Ramankutty et al., 2002). The remaining yield outputs originated from intensiûcation through yield increases per unit land area. However, the regional variation is large, as is the difference between irrigated and rainfed agriculture. In developing countries rainfed grain yields are on average 1.5 t/ha, compared with 3.1 t/ha for irrigated yields (Rosegrant et al., 2002), and increase in production from rainfed agriculture has mainly originated from land expansion.

Trends are clearly different for different regions. With 99% rainfed production of main cereals such as maize, millet and sorghum, the cultivated cereal area in sub-Saharan Africa has doubled since 1960 while the yield per unit of land has been nearly stagnant for these staple crops (FAOSTAT, 2005).

In South Asia, there has been a major shift away from more drought-tolerant, low-yielding crops such as sorghum and millet, while wheat and maize hasapproximately doubled in area since 1961 (FAOSTAT, 2005). During the same period, the yield per unit of land for maize and wheat has more than doubled. For predominantly rainfed systems, maize crops per unit of land have nearly tripled and wheat more than doubled during the same time period.

Rainfed maize yield differs substantially between regions. In Latin America (including the Caribbean) it exceeds 3 t/ha, while in South Asia it is around 2 t/ha and in subSaharan Africa it only just exceeds 1 t/ha. This can be compared with maize yields in the USA or southern Europe, which normally amount to approximately 7–10 t/ha (most maize in these regions is irrigated).

The average regional yield per unit of land for wheat in Latin America (including the Caribbean) and South Asia is similar to the average yield output of 2.5–2.7 t/ha in North America. In comparison, wheat yield in Western Europe is approximately twice as large (5 t/ha), while in sub-Saharan Africa it remains below 2 t/ha. In view of the historic regional difference in development of yields, there appears to exist a signiûcant potential for raised yields in rainfed agriculture, particularly in sub-Saharan Africa and South Asia.

Rural development through sustainable management of land and water resources gives a plausible solution for alleviating rural poverty and improving the livelihoods of the rural poor. In an effective convergence mode for improving the rural livelihoods in the target districts, with watersheds as the operational units, a holistic integrated systems approach by drawing attention to the past experiences, existing opportunities and skills, and supported partnerships can enable change and improve the livelihoods of the rural poor. The well-being of the rural poor depends on fostering their fair and equitable access to productive resources.

The rationale behind convergence through watersheds has been that these watersheds help in 'cross-learning' and drawing on a wide range of experiences from different sectors. A significant conclusion is that there should be a balance between attending to needs and priorities of rural livelihoods and enhancing positive directions of change by building effective and sustainable partnerships.

Based on the experience and performance of the existing integrated community watersheds in different socioeconomic environments, appropriate exit strategies, which include proper sequencing of interventions, building up of financial, technical and organizational capacity of local communities to internalize and sustain interventions, and the requirement for any minimal external technical and organizational support needs to be identified.

Dry Land Farming With Special Reference to Telangana

Dry land farming is an agricultural technique for non-irrigated cultivation of land. It may be defined as: "a practice of growing profitable crops without irrigation in areas which receive an annual rainfall of 500 mm or even less". India has about 108 million hectares of rain fed area which constitutes nearly 75% of the total 143 million hectares of arable land. In India, dry land agriculture accounts for nearly two-thirds of total cropped area and generates nearly half of the total value of agricultural output. Rain fed agriculture encounters several constraints on account of climatic, edaphic, and social factors. Out of the 97 million farm holdings in India, about 76% come under marginal and small categories. The productivity levels of these areas have remained lower across years because of frequent droughts occurring due to high variability in the quantum and distribution of rainfall, poor soil, low fertilizer use, imbalanced fertilization, small farm size and poor mechanization, poor socio-economic conditions and low risk-bearing capacity, low credit availability and infrastructure constraints. Consequently, farmers are distracted from agriculture and tend to migrate to cities to look for alternative jobs. Hence, there is a great need to increase the productivity of rain fed crops and overall net returns to keep the farmers in agriculture. A paradigm shift in rain fed agriculture can be expected through technological thrusts and policy changes.

India has about 47 million hectares of dry lands out of 108 million hectares of total rain fed area. Dry lands contribute 42% of the total food grain production of the country. These areas produce 75% of pulses and more than 90% of sorghum, millet, groundnut and pulses from arid and semi-arid regions. Thus, dry lands and rainfed farming will continue to play a dominant role in agricultural production.

Dry lands, besides being water deficient, are characterized by high evaporation rates, exceptionally high day temperature during summer, low humidity and high run off and soil erosion. The soils of such areas are often found to be saline and low in fertility. As water is the most important factor of crop production, inadequacy and uncertainty of rainfall often cause partial or complete failure of the crops which leads to period of scarcities and famines. Thus the life of both human being and cattle in such areas becomes difficult and insecure.

Despite all improvements in agriculture, we have not yet been able to evolve appropriate practices for our dry land areas. The income of farmers of dry land regions is low. We continue to stress on intensive agriculture on irrigated land but we cannot afford to be complacent with our dry lands. We cannot achieve stability in food production with unstabilized dry land agriculture. Therefore, we are required to adopt improved technology especially developed for dry land agriculture.

The strategies that need to be emphasized:

(1) Land care and soil-quality improvement through conservation agricultural practices, balanced fertilization, harnessing the potential of bio-fertilizers and microorganisms, and carbon sequestration;

(ii) Efficient crops, cropping systems, and best plant types;

(iii) Management of land and water on watershed basis;

(iv) Adoption of a farming-systems approach by diversifying enterprises with high-income modules;

(v) Mechanization for timely agricultural operations and precision agricultural approach;

(vi) Post-harvest, cold-storage, value-addition modules;

(vii) Assured employment and wage system;

(viii) Organic farming;

(ix) Rehabilitation of rain fed wastelands;

(x) Policy changes and other support system; and

(xi) Human-resource development, training and consultancy.

Characteristics of Dry-land Agriculture:

Dry land areas may be characterized by the following features:

1. Uncertain, ill-.distributed and limited annual rainfall;
2. Occurrence of extensive climatic hazards like drought, flood etc;
3. Undulating soil surface;
4. Occurrence of extensive and large holdings;
5. Practice of extensive agriculture i.e. prevalence of mono-cropping etc;
6. Relatively large size of fields;
7. Similarity in types of crops raised by almost all the farmers of a particular region;
8. Very low crop yield;
9. Poor market facility for the produce;
10. Poor economy of the farmers; and
11. Poor health of cattle as well as farmers.

Problems of Dry Land Farming in India

The major problem which the farmers have to face very often is to keep the crop plants alive and to get some economic returns from the crop production. But this single problem is influenced by several factors, they are:

Moisture Stress and Uncertain Rainfall

The rains are very erratic, uncertain and unevenly distributed. Therefore, the agriculture in these areas has become a sort of gamble with the nature and very often the crops have to face climatic hazards. The farmers also take up farming halfheartedly as they are not sure of being able to harvest the crops. Thus, water scarcity becomes a serious bottleneck in dry land agriculture.

Effective Storage of Rain Water

According to characteristics of dry farming, either there will be no rain at all or there will be torrential rain with very high intensity. Thus, in the former case the crops will have to suffer a severe drought and in the latter case they suffer either flood or water logging and they will be spoilt In case of very heavy downpour, the excess water gets lost as run-off which goes to the ponds and ditches etc. This water could be stored for providing life saving or protective irrigation to the crops grown in dry land areas. The loss of water takes place in several ways namely run-off, evaporation, uptake through weeds etc.

Marketing Problem

In dry farming all the farmers grow similar crops which are drought resistant. These crops mature at the same time and the growers like to dispose off their products soon after the harvest. This results in a glut of products in the market and the situation is badly exploited by the grain traders and middlemen. Therefore, marketing becomes a serious problem in dry farming areas.

Unbalance Their Economic Position

Only drought resistant crops namely oilseeds, pulses and coarse grains like jowar, bajra, millets etc. can be grown in dry land areas. Thus, the farmers have to purchase other food grains and household commodities that unbalance their economic position.

Careful and Judicious Manorial Scheduling

In case of irrigated farming the farmers are at a liberty to apply manures and fertilizers according to their availability and facility but in case of dry farming they have to be very careful in fertilizer application. Due to lack of available moisture, broadcasting or top dressing becomes wasteful and meaningless. These can be applied by only deep placement and foliar spray for an improved crop production.

Utilization of Preserved Moisture

Judicious and purposeful utilization of preserved moisture water depends upon soil type, plant type and other factors. The amount of available water to the plants depends upon the depth of plant roots, their proliferation and density. In case of limited moisture condition, the yield directly depends upon the rooting depth. The rooting depth can be desirably increased by mechanical manipulation of the soil. If the planting is very dense and all the plants have same kind of rooting then there will be a tough competition among roots for moisture and scarce moisture condition will result in the wilting of plants. Therefore, utilization of preserved moisture is an art in dry farming. The water collected in ponds or brooks may be used to give protective or life saving irrigation. The widely spaced crops can be intercropped with oilseeds or pulses for increasing the productivity of the land per unit area and per unit time. Therefore, the water collected during the rainy season need special technique and skill for its efficient utilization.

Quality or the Produce

The quality of the produce from dry farming areas is often found to be inferior as the grains are not fully developed or they are not filled properly; often mixed with other crop seeds owing to mixed .cropping system prevalent in these areas and the fodder become more fibrous. All these factors reduce the market value of produce and the farmers do not get the profit of their labour and investment.

Special Reference to Telangana

Telangana, the semi-arid land of India, is experiencing drought often pushing large numbers of people to the margins of living. Drought visits south Telangana "once in two and half years". The rainfall of about 70 cm and less in southern Telangana hardly justifies the fact that the region should languish under semi-arid conditions. In fact, the region forms part of the catchment of the perennial rivers Krishna and Godavari. The irrigation policy initiative over the years continuously favoured the Delta region leaving a large number of people at the mercy of degraded nature and sub-human living. Thus "Telangana backwardness has essentially political roots: with better administration the considerable water resources could have been more fully tapped for irrigation. Telangana is still mainly a dry farming area, the reason for this in Telangana is long term failure to harness the potentialities of the area."

Analysis: As major irrigation facilities are not sufficiently available and short fall of normal rain, when compared to Costal Andhra region, Telangana is mostly dependent on dry land farming. In Telangana, in view of the non-availability of water in wells and tubewells due to depletion of water table and drought farmers are now forced to keep major part of their dry lands as current follow lands. As a result, the area under follow land is increasing year after year. There is a shift in cropping pattern and at the same time, the total area under jowar, bazra, caster and cereals decreased significantly, while the cultivable area under rice, maize, groundnut, oilseeds, cotton, pulses increased proportionately. The yield and production of these crops also increased. It is a definite change in favour of commercial crops.

Due to recurring drought conditions, most of the borrowers in rural areas of Telangana could not repay the loans borrowed earlier. In view of this, financial institutions kept those villages as de-faulted borrowers, included in the black list closing their chance of borrowing again. This has become a stumbling block to majority of the rural households in all the regions in the state, particularly in Telangana. Consequently, the dependency on money lenders and private financiers is again on the increase lending to increase in the cost of production, unremunerative cultivation and increasing indebtedness.

The State needs to give priority for agriculture particularly, in the field of irrigation sector and cheap and assured credit facility. The focus should be on dry land farming, extension services and provision of quality seeds and fertilizers and timely assistance. In recent years the plan allocations to the priority sectors such as agriculture, irrigation have been declining from plan to plan. Irrigation sector is neglected.

Many of the proposed projects in Telangana region could not be undertaken. While total canal irrigation through canals remained stagnant, tank irrigation declined during the last two decades. Similarly, cultivation under dugwells and borewells has increased significantly leading to power problems, and depleting water table below 600 feet in certain areas, and gradual withdrawal of subsidies to agricultural sector also increased cost of cultivation unremunerative cultivation. This has led to unrest among the farmers resulting suicide deaths especially in Telangana region.

Telangana projects have been allocated 266.83 TMC (Thousand Million Cubic Feet. One TMC ft is equivalent to about 28.317 million cubic metres or 22 956.8 acre feet) of water against a due share of 552 TMC. Mahboob Nagar known for its very high levels of distress migration and perennial drought, should have got 187 TMC of water but have received nothing till now. Costal Andhra receives several times more than its due allocation of 99 TMC. Farming has become risky in Telangana, as indicated in the large number of suicides by farmers. Telangana accounts for as many as two- thirds (66%) of the total number of suicides reported in the state between 1998 and 2006. Though recent data shows that Telangana has been allocated a higher share in expenditure on irrigation (55%) than its share of population (41%) however, compared to costal Andhra, the unit cost of irrigation is much higher in Telangana (as it is situated in Deccan plateau) as lifting of water requires huge investments in pumping machinery and power.

As a matter of fact, majority of the households in villages are considered to be labourers. Real development of villages can only be achieved if the labour households' employment, wages and incomes are improved. It is observed that employment, wages and other living conditions of labour households are further deteriorated in recent times. Non-agricultural employment is found to be significant in those areas where canal irrigation is provided. With the development of agriculture, non-agricultural employment was also generated in the command areas. Due to backward agriculture and frequent droughts in most of the mandals of Telangana, labour households find it difficult to get employment during lean seasons and prefer to migrate to far and near places. The process of migration has accelerated in recent years. Due to lopsided developmental strategies pursued from time to time, balanced development of the state has become a casualty and regional imbalances went on widening. These imbalances have become stumbling blocks for the emotional integration of the people of all the three regions of the state. Further, the process of implementation of economic reforms including privatisation is taking place at an accelerated pace in the state.

Impact of Liberalization, Privatization and Globalization

The policies of liberaliation, privatisation and globalization have been displacing the masses from their opportunities. The benefits and subsidies meant for weaker sections are reduced year after year even these meant for the Scheduled Castes and Scheduled Tribes. This is not followed by a corresponding support in alternative occupations or opportunities. With introduction of labour saving technology in the field of construction of roads and buildings, wage labourers have been badly effected in the state. In view of the lopsided pattern of development, the state has been witnessing agitations, movements, rural unrest, farmer suicides and hunger deaths in recent years. The village economy is facing economic and social crisis. Agriculture is unable to absorb the over increasing working population. Further, the cost of production per unit of agricultural output in Andhra Pradesh now is higher compared to major agricultural States in India. The area under canal irrigation system declined due to deceleration in public investment in Telangana. In nutshell, the agricultural sector is neglected by the government. There is a need to review this policy.

Small and marginal farmers have been worst effected. Majority of the small and marginal farmers still depend in informal or non-institutional sources of credit, particularly, money lenders, private financiers at higher rates of interest, consequently, high cost of production and indebtedness.

Liberalisation and Privatisaion process was initiated in the state with firm determination during 1996-'97. But it's impact is not well received by all section of the people. Agricultural growth rates have gone down drastically. Employment situation in rural areas were not improved rather deteriorated. Whatever the employment opportunities have been created so far, they are largely low paid and casual in nature and insecure. Nonagricultural employment could not be generated to the levels of expectations. Villages have become markets for products of multinational, and big industries. Whatever the industries, or small scale industrial units were available earlier, they are unable to compete with global products either in quality or prices. Backward area like Telangana (except Hyderabad) could not attract either domestic or foreign direct investment.

There is an exodus of young persons from villages of backward and drought effected districts to towns and cities in search of livelihood. People from Mahaboobnagar, Nalgonda, Karimnagar, Warangal and Medak are migrating to gulf countries and Hyderabad to get some livelihood or other. Only old age people keep staying in rural areas. The Information and Technology could provide jobs to a few thousands of educated young people.

The Self-help Groups for women could not provide work as expected. This programme could enlighten rural women groups in political and social aspects. These groups could mobilise savings out of their hard earned income besides State/ Central assistance. As far as employment and income generation activity of this programme is concerned, very little is achieved. Whatever the products are produced by these groups, they are decorative and artistic, unable to compete with global multinational products. Mostly they are neither mass consumption oriented nor essentials. Hence, they suffer from lack of demand. The scheme has become political wing of ruling party for vote bank.

Suggestions:

(1) Agriculture may be given top priority along with infrastructure development in backward regions. Constructions of Ichampally irrigation project across river Godavari will benefit north Telangana. Similarly, through proper allocation and utilisation of Krishna river water will also benefit Nalgonda, Mahabubnagar districts in South Telangana.

(2) Distribution of cultivable public lands surplus lands and cultivable waste lands among the rural poor provides some solution to the agricultural labourers.

(3) There is an urgent need to change the cropping pattern in drought prone areas of the regions to prevent further downslide of under ground water table.

(4) It is also necessary to identify backward districts and specific area programmes may be initiated through state and central grants.

(5) Rural and agricultural credit facilities have to be adequately provided to all the needy households keeping in view the growing dependency of farmers and rural artisans on money lenders and private financiers.

(6) Both central financial transfers and use of policy instruments will be useful to attract private investment to the backward regions; and

(7) Enhanced allocations for social development such as education, health, nutrition, empowerment of poor. Further democratization of rural institutions etc., will improve education, skills and entrepreneurial abilities of people in backward areas.

Conclusion: Even after utilizing all the available water resources, about 50% of our cultivable area will still depend on rains. Therefore, our agricultural scientists, policy formulators and farmers should appropriately realize the magnitude of role that. rainfed agriculture or dryland farming can play. They should thoroughly examine the problems of dry land agriculture from different viewpoints and evolve appropriate technologies, crop varieties, etc. for these areas to better the economic position of the farmers. Dry farming areas, therefore, need a much closer attention for achieving food security in India.

Problems of Crop Production in Dryland

Dry farming crops are characterized by very low and highly variable and uncertain yields. Crop failures are quite common. These are mainly due to the following causes.

Inadequate and Uneven Distribution of Rainfall

In general, the rainfall is low and highly variable which results in uncertain crop yields. Besides its uncertainty, the distribution of rainfall during the crop period is uneven, receiving high amount of rain, when it is not needed and lack of it when crop needs it.

Late Onset and Early Cessation of Rains

Due to late onset of monsoon, the sowing of crop are delayed resulting in poor yields. Sometimes the rain may cease very early in the season exposing the crop to drought during flowering and maturity stages which reduces the crop yields considerably

Prolonged Dry Spells During the Crop Period

Long breaks in the rainy season is an important feature of Indian monsoon. These intervening dry spells when prolonged during crop period reduces crop growth and yield and when unduly prolonged crops fail.

Low Moisture Retention Capacity

The crops raised on red soils, and coarse textured soil suffer due to lack of moisture whenever prolonged dry spells occur due to their low moisture holding capacity. Loss of rain occurs as runoff due to undulating and sloppy soils.

Low Fertility of Soils

Drylands are not only thirsty, but also hungry too. Soil fertility has to be increased, but there is limited scope for extensive use of chemical fertilizers due to lack of adequate soil moisture.

Agrodiversity as a Means of Sustaining Small Scale Dryland Farming Systems in Tanzania

The Issues

Dryland farming systems in Africa are often characterised as being extremely degraded, vulnerable to external forces, and low in productive output. Available fossil evidence, for example, suggests that the open savannas of East Africa have the longest history of human habitation since Pliocene times. As a result, Adams (1996, p.208) observes that, "the destructive impact of long human occupation of the ecosystem has been profound. ..... Man's tendency to overexploit the basis of his subsistence is endemic." Such stereotypes of land use abound. The story of man's mistreatment of his environment has been recounted many times and continues to be told. Lord Hailey (1938) described human-induced soil erosion as "the scourge of Africa", while the FAO (1990, p.6) bemoans that, "Africa's lands are under attack." This paper targets a very different tendency – human's propensity to conserve. Some land uses, developed over centuries of pressures and difficulties, display a resilience and productivity that is truly remarkable. Both professionals and policy makers may draw some comfort that stereotypes of environmental crisis are not applicable throughout dryland Africa.

Nevertheless, environmental change is occurring. Global biodiversity loss in areas of land use is a well-attested phenomenon. Ecologists, in particular, are alarmed at how natural biological diversity is being replaced by relative biological uniformity, especially under the pressure of population growth (Cincotta and Engelman, 2000). As natural habitats decline, greater proportions of species living within those habitats become extinct. Species-area curves suggest that at about 10 percent of land area devoted to protection, only 45 to 70 percent of species remain, and as habitat declines further, extinctions accelerate dramatically (Pimm *et al*, 1995). Only Costa Rica comes anywhere near the conservationist's goal of maintaining at least 10 percent of land areas under natural habitat. However, there is good evidence that natural biological diversity may be giving way to another diversity, equally valuable and of greater immediate significance to society, which in this paper is called 'agrodiversity'. It embodies cultural and spiritual dimensions of biodiversity (Posey, 1999), as well as practical and economic values of gaining a sustainable rural livelihood for poor people (Altieri, 1999). Without in any way denying the need to conserve natural biological diversity upon which ecologists argue many of the world's life-support functions are based, this paper dwells on the benefits of 'agrodiversity' and empirical examples from Tanzania on its value to local society and, indeed, to the whole conservation agenda.

Introducing Agrodiversity

Awareness of global biodiversity loss has resulted in a search for where the battle-lines for its protection should best be drawn. Typically, these lines have been fortified around the various types

of protected areas and controlled management of habitats. Increasingly, this 'fortress conservation' has become unsustainable in the face of social change and population pressures (Ghimire and Pimbert (1997). Additionally, although evidence for the relative numbers of species is elusive, there is likely to be far more biodiversity in areas of land use than in all protected areas together, principally because areas that are used are not only far greater than those protected but also these areas are the more fertile and naturally biodiverse. Consequently, it makes sense for the global biodiversity agenda to look to:

- natural biodiversity managed within farming systems, including domestication of wild species and protection of rare species;
- the farming systems themselves, including the way that biodiversity is managed for the welfare of land users such as techniques and practices of cropping, agroforestry and rangeland management;
- how biodiversity itself helps vulnerable farming groups cope with social, economic, political and demographic pressures as well as variable and marginal physical environments.

While the term 'agricultural biodiversity' or 'agro-biodiversity' has tended to be used to signify the variety of plants, species and varieties on the lands of farmers and their relation to welfare (Thrupp, 1998), 'agrodiversity' is much more broadly defined. It encompasses "the many ways in which farmers use the natural diversity of the environment for production, including .... their choice of crops....[and] their management of land, water, and biota" (Brookfield and Padoch, 1994, p.9). It goes beyond the concept of species and genetic diversity of plants and animals to incorporate other aspects of the farming system that relate to obtaining sustainable livelihoods.

Because small-scale farmers in the tropics usually have to rely on the intrinsic quality of their natural resources, including biodiversity and soil quality, and because they have limited resources to invest in external inputs, 'agrodiversity' is a crucial underpinning to their lives.

This paper, therefore, illustrates some aspects of the importance of agrodiversity to the ways in which small-scale farmers in a dryland part of Tanzania cope with environmental and social pressures. It takes up Paul Richards' (1985, p.160) challenge that, "the evidence .... of innovativeness in the peasant food crop sector is strong, but further work is needed on ...how....this pool of skill and initiative might be harnessed to national development objectives."

The policy implications of agrodiversity will be developed in the last section of this paper. First, however, we describe the approach of the GEF-funded project, *People, Land Management and Environmental Change*, which is compiling a database of agrodiversity in order to demonstrate its value and potential.

PLEC Approach

The aim of the People, Land Management and Environmental Change (PLEC) project is to develop sustainable and participatory approaches to conservation within small farmers' agricultural systems, and in participation with farmers.

The specific objectives of PLEC are:

a) to establish historical and baseline comparative information on agrodiversity, including biodiversity, at the landscape level;
b) to develop participatory and sustainable models of biodiversity management based on farmers' technologies and knowledge within agricultural systems at the community and small-area levels;

c) to recommend approaches and policies for sustainable agrodiversity management to key government decision makers, farmers, and field practitioners; and

d) to establish national and regional networks for capacity strengthening within participating institutions, and to carry forward the aims of PLEC.

The core of PLEC's work is in its 'demonstration site' villages. Here, PLEC becomes the farmers' own enterprise, and scientists are the facilitators, not the instructors. The scientists identify and demonstrate farmers' practices that are environmentally, socially and financially sustainable, and which sustain biodiversity. They help farmers in achieving their own conservationist goals. Collaborating farmers manage varied biophysical conditions, growing a range of crops and using biodiversity with discretion. The PLEC approach differs from mainstream agricultural research at experiment stations under controlled conditions. By integrating locally developed knowledge of soil, climate, and other physical factors with scientific assessments of their quality in relation to crop production, a set of sustainable agricultural technologies can be devised so that agricultural diversity is maintained. The participatory process will eventually enhance farmers' and local communities' ability to adapt to environmental, social and economic change.

The PLEC project, through demonstration sites and farmer-to-farmer extension, seeks to support existing diversity and disseminate it to other farmers whose positions might be improved as a result of new practices or techniques. PLEC relies on interaction between farmers and between farmers and scientists as a mechanism for identifying and maximising agrodiversity. The Tanzanian element of the East African PLEC cluster focuses on two contrasting demonstration sites in Arusha region in Northern Tanzania. The characteristics of each are summarised in the following table.

Table: *Characteristics of PLEC sites in Northern Tanzania.*

Characteristic	***Ngiresi/Olgilai***	***Kiserian***
Average altitude	1,900m asl	1,200m asl
Temperature range	12-30° C	12-30° C
Mean annual rainfall	2,000mm	500mm
Farming system	Mixed cropping with zero grazing	Agropastoral
Soil type	Andosols	Cambisols
Village population (1988)	2,158	3,330

In Tanzania the PLEC scientists initially met with farmers to identify environmental and social constraints and to see how coping strategies were related to those constraints. Following on from this, 'expert' farmers were identified and, using PRA techniques and training sessions, linkages were established with other farmers. The focus has been on identifying ways in which farmers have adapted their practices to, and have made use of, the environment in which they farm while at the same time conserving or enhancing agrodiversity, especially biodiversity.

Diversity in Crops and Cropping Systems

Planting a mix of crops is recognised as a way of spreading risk on a farm. Within crop groups, different varieties are planted to match the particular stresses of the local environment. Farmers seek to plant a combination of varieties that will ensure at least some yield despite extremes of climate, pests, disease, labour shortage or other constraints.

In a similar way, farmers select the crop types that they plant to enhance their food security. For example, in Ngiresi sweet potatoes are planted on steep slopes not only to provide cover and thereby reduce erosion, but also because they mature in the 'hungry period' before maize is ready for harvest. In addition, sweet potatoes will produce even in a drought year.

Farmers, therefore, select crops and varieties using different criteria – some strains will be selected because they are high yielding in optimum conditions, others because they are tolerant to drought, others because they are resistant to storage pests, have a high market price, good taste or are easily processed. Each selection involves an assessment of the potential risks and rewards of planting a particular crop or variety. Such decisions are influenced not only by the physical characteristics of the environment, but also by socio-economic factors such as available labour and proximity to markets.

Farmers view the selection of crop varieties as a continuous process. Some varieties that are tried in the field become part of the farmer's own landrace, whereas others, whose characteristics prove to be less suited to the local environment, quickly disappear from the field. Crops and varieties that continue are those identified as being a best match for the field conditions, the wider environment and the farmer's own situation.

In the Tanzanian PLEC sites maize, beans and bananas are very widely grown. Within these crop types a wide spectrum of varieties is planted. The following table lists some of the varieties identified on farms. This table clearly illustrates that, while research stations may breed for high yields, drought tolerance and pest resistance, farmers consider a wide range of other characteristics before selecting which seeds to plant.

Issues such as intercropping compatibility and labour availability are also relevant to farmer decision-making. So for example, trailing beans are less popular than the bush varieties because they get entangled with the maize intercrop. More labour is required to disentangle them and beans are lost when pods burst during this process. Most farmers plant up to five different varieties of beans and bananas on their land.

This reflects not only risk-spreading decisions, but also the matching of particular varieties to scattered plots having different biophysical characteristics. Traditionally, maize and beans have been intercropped in the two sites. No standard proportions are used, rather the exact combinations grown depend on assessments by individual farmers of the specific plot, previous experience on that plot, market conditions, location and topography of the field etc. Typically, in Kiserian, a higher proportion of beans are planted (2 rows of beans for every row of maize). This reflects not only the fact that beans are a highly valued food crop in their own right, but also the higher market prices that can be achieved for beans over maize and bean crop residues over maize stalks (as livestock fodder). The greater susceptibility of beans to pests and diseases is also cited as another reason why farmers plant more beans, so that they can be guaranteed some yield from this important crop.

***Table:** Maize, Bean and Banana Varieties Grown in Arumeru*

Crop varieties	*Economic uses*	*Plant characteristics*
Zea mays (Maize)		
Kienyeji	Food, income, crop residuesfed to animals	Not very sweet, tolerant to storage pests, good milling quality, low yielding, drought susceptible.
Katumani	Food, income, crop residues fed to animals	Drought tolerant, early maturing, low yielding, good milling quality, tolerant to storage pests.
CG4141 (Lowlands)	Food, income, crop residues fed to animals	Good milling quality, drought tolerant.
UCA (Highlands)	Food, income, crop residues fed to animals	Good milling quality, drought tolerant.
Kilima	Food, income, crop residues fed to animals	High yielding, susceptible to storage pests, good milling quality, high water demand, high quality flour.

Phaseolus spp (Beans):		
Soya kijivu	Food, income, crop residues fed to animals.	"No gases after eating", early maturing, good taste, climbing type, sweet, high price, grey.
Kachina	Food, income, crop residues fed to animals.	High market price, early maturing, spoils quickly after cooking.
Lovirondo	Food and crop residues fed to animals	Climbing type, "causes bloating and gases after eating", laborious to harvest, low market price.
Bwanashamba	Food and crop residues fed to animals	Most popular in Kiserian, high yielding, good taste, susceptible to diseases and aphids.
Masai red ndogo (namira)	Food and crop residues fed to animals.	High yielding, good tasting, "no gases after eating", needs wide spacing for high production.
Karanga	Food and crop residues fed to animals.	High yielding, good tasting when cooked (flavours food).
Masai-red kubwa (namriri)	Food and crop residues fed to animals.	High market price, bush type, early maturing, good tasting and flavours food, susceptible to diseases.
Lyamungu 90	Food and income	Good tasting and flavours food, early maturing, drought tolerant, high yielding, high market price.
Kiburu	Food and crop residues fed to animals.	Drought tolerant, grows well on soils with poor fertility.
Engichumba	Food and income	Very high yielding, violet bean
Engichumba-ng'iro	Loshoro (traditional food)	High yielding, sweet, grey bean
Engichumba-narok	Food and income	Similar to Engichumba-ng'iro, black bean
Moshi	Food and income	Very high yielding, sweetest, yellow bean
Kibumulu	Food and income	Fast cooking, high price, dark red bean
Musa spp (Bananas)		
Kisimiti	Income, brewing, animal feed (stem)	Early maturing, drought tolerant, good milling quality
Ng'ombe	Loshoro, brewing, income, roofing, fodder to animals.	Hard when cooked.
Mshale	Matendela (traditional food), income	Good for roasting, long and thick banana fingers.
Uganda fupi	Banana soup (mtori), fruit, income, peels fed to animals	Early maturing, small with mainly fingers, susceptible to pests and diseases.
Uganda ndefu	Banana soup, fruit, peels fed to animals	Large with few fingers, susceptible to pests and diseases.
Kisukari	Fruit, income, animal feed (stems)	Very sweet, drought and disease tolerant, low nutrient demand
Mzuzu	Roasting for tea	Tolerant to drought and disease
Malindi	Food (matendela), animal feed	Drought tolerant

Mnanambu	Soup, roasting	Shade
Mkonosi	Roasting	Disease tolerant
Mkono wa tembo	Roasting	Disease tolerant
Ndishi	Loshoro, income	Susceptible to diseases
Olmuririko	Loshoro, brewing	Modest tolerance to diseases

Diversity at Landscape Level

It is typical in Ngiresi/Olgilai for farms to be made up of a number of different plots scattered throughout the village. The evolution of these farm types is explained first by the periodic clearance of more land as additional areas were brought into production and second by the practice of dividing farms between sons. In such divisions parents are inclined to split every plot into smaller plots to ensure that all sons receive land of similar quality. Farmers in Ngiresi and Olgilai use different parts of the landscape in different ways, matching crop suitability to available land. In matching crops to plots of land, farmers not only consider biological suitability, but also the value of the crop, the labour required to manage the crop and the risk that the crop might fail. Therefore, high value crops and fruit trees are planted close to the home to reduce the risk of theft, whereas hillside plots are planted with sweet potatoes to reduce soil loss from erosion. Field borders are often planted with bananas and trees to demarcate boundaries. Farmers recognise the suitability of different plots to different cropping strategies. For example, Mr. Yangan from Ngiresi/Olgilai, has nine different field types making up his farm. Some of these are detailed below.

***Table:** Some Field Types on Mr. Yangan's Farm*

Location	*Field Type*	*Rationale*
Near house flat/gentle slopes	Tethering pasture	Convenient for milking morning and evening.
	House garden	Convenient to house
	Coffee/banana	Allows for easy transport of manure to the field. Also convenient for regular chores such as spraying coffee.
Middle distance gentle slopes	Maize/beans	Convenient enough for transporting inorganic fertilizers. In situ green manuring replaces animal manure.
Remoter steep slopes	Sweet potatoes Fodder grasses	Crops don't need a lot of inputs, but provide good cover for hillsides prone to erosion. Sweet potatoes in particular attract a lot of mice so it is preferred that these are not grown close to the house.
Boundaries	Bananas	Tree crops in particular are used to mark out boundaries.

Some of the older farmers in Kiserian have a deep understanding of the different soil types found in the village. Soils are classified locally by colour, fertility, depth and moisture holding capacity. Based on this categorisation crops are selected to match different soils and planting dates are determined. In order to spread risk some plots and crops are dry planted, others are planted immediately after rain, while still others are planted a few days after the first rains. In these ways farmers are able to minimise the risks of poor crop yields due to low levels of soil fertility and moisture stress.

Diversity in Resource Management

Management practices reflect the value placed on the land and on the current crop by a farmer. Land tenure is often critical in determining how farmers manage their land. Secure tenure has been shown to be an incentive to careful stewardship of the land. Open access property or fields held on short term rents are often less well tended. In the case of Kiserian the contrast in management practice is evident in the condition of communal and clan woodlots compared to those owned individually. The former are degraded while the latter are actively managed by the owner (for example, through the application of crop residues).

Table: *Soil management strategies under the major cropping systems in Olgilai/Ngiresi villages.*

Management Objective	*Cropping system*	*Soil Management Strategies*
Soil fertility improvement	Coffee/banana/maize/beans in rotation with round potatoes	FYM application, incorporation of crop residues, house refuse and weeds and application of ashes, planting of Sesbania sesban, grevillea and composting (few).
	Maize/beans	Incorporation of FYM, grevillea biomass, crop residues, green manuring, trash lines.
	Maize/beans rotation with sweet potatoes	Application of FYM, incorporation of crop residues and Sesbania biomass.
Soil moisture management	Coffee/banana/maize/beans in rotation with round potatoes	Incorporation of crop residues, mulching and trash lines, shading by coffee/banana/fruit trees/ canopy, planting of seteria.
	Maize/beans	Self-mulching, crop residues, trash lines, incorporation of crop residues (few).
	Maize/beans rotation with sweet potatoes	In-situ crop residues mulching, trash lines, sweet potato cover.
Soil erosion control	Coffee/banana/maize/beans in rotation with round potatoes	Rain interception by trees canopy, Mulching, trash lines, construction of flower hedges, trees canopy interception, planting of Sesbania sesban.
	Maize/beans	Trash lines, crop canopy
	Maize/beans rotation with sweet potatoes	Incorporation of crop residues, trash lines, application of ashes, planting of fodder grass strips.

Management practices also vary with landscape. Fertility management, in particular, differs depending on the spatial arrangement of a farmer's fields. In general, farmyard manure will be applied to fields close to the house/cow stall. Fewer farmers are likely to transport manure long distances for application on hillside plots. Instead they may use other biological means of enhancing soil fertility, or they may apply chemical fertilizers, which are easier to transport.

The following table details some of the soil management practices that are applied by farmers to fields in Olgilai/Ngiresi

In general, there is less emphasis in Kiserian on soil fertility management, with less use of both organic and inorganic fertilizers. Instead, in order to increase crop yields, farmers rely on bringing more land into cultivation, mixed cropping and variations in planting dates.

Agrodiversity – Expert Farmers

The PLEC approach is to identify 'expert' farmers and to facilitate the dissemination by them of their knowledge and experience. In this context expert farmers are those who have maximised the production potential on their farms in a sustainable and conserving way. In both Ngiresi/Olgilai and Kiserian this involves diversification and intensification of crop production.

Gidiel Laizer farms six separate plots in Ngiresi/Olgilai. He plants different crops on different plots depending on their distance from his home, their topography and the other resources available to him (especially labour).

In 1998 Gidiel changed his planting strategy on a 0.25 ha plot surrounding his house. He now continuously crops this field. This plot is planted with the traditional perennial intercrop of coffee and bananas. However, Gidiel no longer plants the usual intercrop of maize and beans.

Instead, in response to the availability of local markets (Arusha town is just a few kilometres away), he plants cauliflower and round potatoes for sale in addition to maize for home consumption.

This serial cropping system, yielding three crops per annum, is called a 'matatu' system and involves continuous cultivation of the plot. In the following table Gidiel's planting and harvesting calendar is summarised.

Table: *Matatu cropping system*

Month	***Activity***		
	Planting	***Harvesting***	***Other***
March	***Plant cauliflower***		***Manure applied***
April			
May	Plant round potatoes	Harvest cauliflower	
June			
July		Harvest round potatoes Harvest bananas	Manure applied
August	Plant maize	Harvest bananas	
Septembe r			
October			
Novembe r		Harvest coffee	Manure applied
Decembe r		Harvest coffee	
January			
February		Harvest maize	

In addition, further crops are planted along the plot boundary providing further sources of food which include taro (*Colocasia esculenta*), fodder (*Cariandra caryopsis*) and grevillea (*Grevillea robusta)* for which he sells timber to get cash. Gidiel is also planning to introduce yams (*Dioscorea spp*) into the border of the field as an additional source of food. High levels of diversity are supported on Gidiel's farm. In addition to the 'matatu' cropping system, which supports diverse crops in a single field, varietal diversity is also evident. Gidiel cultivates five different varieties of bananas and he is attempting to crossbreed two varieties of maize to develop his own strain combining the desirable traits from those that he is currently growing.

Gidiel selectively applies farmyard manure to the plots close to the house where he practises the 'matatu' system. He identifies those areas of the field where the yield was poor for the previous crop. Those parts of the field receive the first and greatest application of manure in order to improve their productivity under the next crop.

As a result of health concerns and for financial reasons, Gidiel does not use commercially produced chemical pesticides.

Instead he uses a mixture of fermented cattle urine (collected from the stall) and water, which he prepares himself.

Spraying this mixture on the plants also introduces additional nutrients to the crop. Another use of fermented urine (mixed with botanicals) is to control ticks in livestock.

The PLEC approach seeks to encourage the communication of information by farmers to farmers.

This is facilitated by farmer demonstrations, where farmers are invited to visit the farm of an 'expert' farmer who explains his practices.

In addition to the innovative matatu cropping system, Gidiel cultivates a large number of plants with traditional medicinal uses. Following demonstration meetings held at his farm, local farmers took away cuttings and seedlings from plants used in traditional medicine to propagate them on their own farms.

In Kiserian, the strategies followed by Lais Kitia, another expert farmer, differ from those of Gidiel Laizer, largely due to the different environments in which they farm.

Lais Kitia has established, over many years, a very successful agroforestry cropping system. In addition to cereals and legumes, Lais maintains a wide variety of useful trees, many of them local species, well adapted to the arid conditions. His agroforestry and cropping systems mean that he harvests a crop, be it cereal, beans or fruit, every few months.

The fruits are mostly harvested during the dry season and provide an important source of food for his family.

Because Kiserian is more remote from markets, Lais's production is focused on foodstuffs for home use rather than for sale in the market, although produce from his agroforestry system can be converted into cash if necessary.

Lais's farm has ten diverse field types. The crops grown in a single plot of less than 0.3ha are summarised in the following table. In addition to two annual crops (maize and millet), this field supports a wide variety of fruit trees and is used for tethering cattle and sheep.

Table: *Farming on One Field at Mr. Lais Kitia's Farm in Semi-arid Kiserian*

Type of plant	*Varieties*	*Economic uses*	*Plant characteristics*
Maize (*Zea mays*)	Malawi	Food, income, animal feed, firewood	High yielding, large grains, easy to mix with trees, good milling qualities.
Finger millet (*Eleusine indica*)	Enyangai (local)	Porridge, income, brewing, animal feed	High price, tolerant to drought and storage pests.
Mangoes (*Mangifera indica*)	Embedodo	Fruit, income	Large fruit, sweet, good aroma, good market price, keeps, green when ripe, drought tolerant.
	Embe mviringo	Fruit, income	Very sweet, fibrous, heavy, green when ripe, drought tolerant.
	Saforoni	Fruit, income	As sweet as honey, drought tolerant.
	Boribo	Fruit, income	Very sweet, high water content, orange and red when ripe, drought tolerant.
	Achari	Fruit, income, combines with chilli pepper, tomatoes, onion and coconut juice to make appetiser.	Very sweet, drought tolerant.
Pawpaws (*Carica papaya*)	Kienyeji 1	Fruit, income	Watery, very sweet, soft.
	Kienyeji 2	Fruit, income	Moderate water content, moderately sweet, hard.
Lemons (*Citrus limoni*)	Kienyeji	Income, porridge appetiser	Very sweet, drought tolerant.
Zambarau (*Syzygium guineense*)	Local	Fruit, income, shade.	Drought tolerant
Oranges (*Citrus cinensis*)	Local	Fruit, shade	Drought tolerant
Ukwaju (*Tamarindus indica*)	Local	Wild fruit, income	Drought tolerant, soil fertility improvement
Bananas (*Musa sapientum*)	Kisukari	Fruit, income, fed to animals.	Very small fingers, very sweet, drought tolerant.
	Kisimiti	Fruit, income, brewing, fed to animals.	Drought and disease tolerant
Mnafu (*Solanum nigrum*) (volunteer crop)	Local	Wild vegetable	Drought tolerant.
Wild amaranthus (*Amaranthus thunbergii*) (volunteer crop)	Local	Wild vegetable	Drought tolerant
Livestock	Varieties	Economic uses	Characteristics
Cattle (tethered on wild trees)	Local breed	Draught power, milk, manure, income, security, dowry, prestige.	Pests and diseases tolerant.
Sheep (tethered on wild trees)	Local	Meat, dowry, fatty foods to breast feeders.	Pests and diseases tolerant.

Conclusion and Policy Implications

Dryland farming systems in Africa display a remarkable resilience. If the past predictions of their demise were to have been true, we should be seeing far more environmental degradation and loss of biodiversity than we do today. Part of the reason for the resilience lies in the ability of farmers to adapt to changing conditions. The Boserupian hypothesis of adaptive change through the local application of technology and sustainable intensification is alive and well in many, but obviously not all, parts of Africa. This paper has sought to show that farmers in Arumeru, Tanzania, are no exception. They have developed some sometimes-intricate sets of techniques to manage their natural resources, they have used the biological diversity they have to hand, especially in regard to food crops, and they have organised their use of the whole landscape to secure their livelihoods and biodiversity.

Of particular relevance to developing lessons from these empirical findings are the ways in which farmers both understand and manage complex associations of plants, which then in turn provide food security to local households. Scientists and policy makers must be very careful not to write off the value in current practices of small-scale farmers as being of no relevance to today.

Debates about the unsustainability of dryland farming in Africa are rife with accusations about the degrading practices of small farmers. However, if land users have themselves developed sustainable and productive practices, then this is as worthy of publicity as the examples of bad practices. Bad news may sell newspapers; but good news should surely be equally prominent in giving policy makers a balanced picture.

The Arumeru farmers have developed an extraordinary diversity in crops and cropping systems. The PLEC project will be looking into what resources farmers need to access this rich harvest of information. However, it is already clear that poor farmers can, and do, practice sustainable use of biodiversity, and many use genotypes, varieties and landraces of crops that are now rare.

Food security for the many marginal populations in Africa, a crucial area of policy debate today, may well rest on these poor farmers' efforts. The Kiserian semi-arid site, for example, has bean varieties that withstand drought as well as provide both food to humans and crop residues to animals.

Multi-purpose varieties and plants that meet several needs are important for dryland farmers. If they also avoid – or minimise – the inherent riskiness of the climate, such plants are worthy of conservation. They may not yield as highly as improved varieties, but for food security at household level this is not the key criterion.

The PLEC experience in Tanzania has also highlighted the beneficial, mutual interaction between professionals, field workers and farmers. Working closely with farmers on demonstration sites, our scientists have come to appreciate that they each have different kinds of knowledge, which are of equal importance. The "white-coat" syndrome that all that comes from scientific experimentation is superior – and by extension, all that comes from small-scale farmers and peasants is inferior – has bedevilled science. Our colleagues in PLEC have learnt to harmonise their experience with farmers and to attempt to interleave their knowledge from science with farmers' own experiences and advice. It has been a good mutual learning exercise for all involved. Policies on extension, agricultural education and interactions with clients (i.e. farmers) have much to gain.

Finally, PLEC in Tanzania has demonstrated the heterogeneity in rural society. There are innovative and expert farmers who are carrying on agricultural activities that have substantial scope for replication. These self-same farmers have training abilities also. Farmer-to-farmer extension has been a particular success in Arumeru, where field days and farmer-organised demonstrations have influenced many land users.

We have witnessed substantial change through such informal means of dissemination. We have seen women farmers teaching each other, swapping planting materials, and enthusiastically engaging in community works. By formally valuing the human resources at local level and appreciating the agrodiversity farmers have developed themselves, PLEC is empowering people and giving them a dignity that more traditional projects fail to emulate. Time will tell if the process will continue. Much will depend on political stability, social order and the continuing willingness of scientists and land users to collaborate. However, the signs are positive. Agrodiversity can help to support the global agenda of conservation of biodiversity, while at the same time providing for development of local people and the meeting of their needs.

Monsoon

A Monsoon of the Indian subcontinent is among several geographically distributed observations of global monsoon taking place in the Indian subcontinent. In the subcontinent, it is one of oldest weather observations, an economically important weather pattern and the most anticipated weather event and unique weather phenomenon. Yet it is only partially understood and notoriously difficult to predict. Several theories have been proposed explaining the origin, the process, the strength, the variability, the distribution and the general vagaries of the monsoon of the Indian subcontinent, but understanding of the phenomenon and its predictability are still evolving. The unique geographical features of the subcontinent, along with associated atmospheric, oceanic and geophysical components, are extremely influential in ensuring the anticipated behaviour for a monsoon in South Asia and the Indian subcontinent. Due to its effect on agriculture, flora and fauna and the general weather of India, Bangladesh, Pakistan, Sri Lanka, etc., among other economic, social and environmental effects, a monsoon is one of the most anticipated, followed and studied weather phenomena of the Indian subcontinent.

It has significant impact on the overall well-being of subcontinent residents and has even been dubbed the "real finance minister of India".

Definition

Monsoon, derived from the Arabic word "Mawsim" meaning "season", although generally defined as a system of winds characterized by a seasonal reversal of its direction, lacks a consistent detailed definition. Some examples are given below:

- The American Meteorological Society defines it as *a name for seasonal winds*, first applied to the winds blowing over the Arabian Sea from northeast for six months and southwest for six months. Later it has been extended to similar winds in other parts of the world.
- Intergovernmental Panel on Climate Change (IPCC) describes Monsoon as a tropical and subtropical seasonal reversal in both the surface winds and associated precipitation, caused by differential heating between a continental-scale land mass and the adjacent ocean.
- Indian Meteorological Department defines it as the seasonal reversals of the wind direction along the shores of the Indian Ocean, especially in the Arabian Sea, that blow from the southwest during one half of the year and from the northeast during the other half.
- Colin Stokes Ramage in *Monsoon Meteorology*, International Geophysics Series, Vol. 15, defines Monsoon as a seasonal reversing wind accompanied by corresponding changes in precipitation.

Monsoon of the subcontinent is primarily noted for its rain bearing ability and for the associated unpredictability of the weather. Consequently some definitions incorporate rain in its definition.

Background

Observed initially by sailors in the Arabian sea travelling between Africa, India and South-East Asia, Monsoon is a major weather phenomenon in India (and the subcontinent) for the influence it casts on the lives of its inhabitants for centuries. Monsoon in India can be categorized into *two branches* based on their spatial spread over the sub-continent:

- Arabian Sea Branch
- Bay of Bengal Branch

Figure: *SW Monsoon clouds over Tamil Nadu*

Alternatively, it can be categorized into *two segments* based on the direction of rain bearing winds:

- South-West Monsoon (SW Monsoon)
- North-East Monsoon (NE Monsoon)

Based on the time of the year that these winds bring rain to India, they can also be categorised in *two rain periods* called:

- the *Summer monsoon* (May to September)
- the *Winter monsoon*, (October to November)

The complexity of Monsoon as a weather phenomenon of India is not yet completely understood, making it difficult to accurately predict its behaviour in terms of quantity, temporal and spatial distribution of the accompanying precipitation. These are the most monitored components of Monsoon determining the water availability in India for any given year.

Mechanism of Monsoon

Monsoon is a tropical phenomenon. Indian subcontinent, lies northwards of the equator up to the Himalayas and Hindukush, primarily in the tropical zone of the Northern Hemisphere. Weather pattern involves winds blowing from the south-west direction (known as South-West Monsoon) from the Indian Ocean onto the Indian landmass during the months of June through September. These are generally rain-bearing winds, blowing from sea to land, and bring rains to most parts of the subcontinent.

They split into two branches, the *Arabian Sea Branch* and the *Bay of Bengal Branch* near the southernmost end of the Indian Peninsula. They are eagerly awaited in most parts of India for their agricultural and economic importance.

Subsequently later in the year, around October, these winds *reverse* direction and start blowing from a north-easterly direction. Given their land to sea flow, from subcontinent onto the Indian Ocean, the system with less moisture brings rain to only limited parts of India like Kerala, Andhra Pradesh and Tamil Nadu.

This is known as the North-East Monsoon. However, this rain is responsible for the rice bowls of South India. This *mechanism* completes the *annual Monsoon cycle* of the Indian subcontinent.

Effect of Geographical Relief Features

Although the SW and NE Monsoon winds are seasonally reversible, they do not cause precipitation on their own.

Two factors for rains are essential for rain formation:

1. Moisture-laden winds
2. Droplet formation

Additionally, one of the causes of rain need to happen, which in this case is primarily Orographic due to presence of highlands right across the paths of the winds.

Orographic barriers in the path of a wind force it to rise. Consequently, precipitation occurs on the windward side of highlands due to adiabatic cooling and condensation of the rising motion of the moist air.

For all the above scenarios to fulfill simultaneously, the unique geographic relief features of Indian subcontinent come into play. The notable features of Indian sub-continent, required in explanation of the Monsoon mechanism, are as follows:

1. First is the presence of abundant water bodies around the subcontinent - Arabian Sea, Bay of Bengal and Indian Ocean. These help in accumulation of moisture in the winds during the hot season.

2. Second is the presence of abundant highlands like the Western ghats and the Himalayas right across the path of the SW Monsoon winds. These are the main cause of the substantial orographic precipitation all over the Indian subcontinent.
3. The Western Ghats are the first highlands of India that the SW Monsoon winds encounter. The Western Ghats rise very abruptly from the Western Coastal Plains of the subcontinent making effective orographic barriers for the Monsoon winds.
4. The Himalayas play more than the role of just the orographic barriers for Monsoon. They help in its confinement onto the subcontinent. Without it, the SW Monsoon winds would blow right over the Indian subcontinent into China, Afghanistan and Russia without causing any rain.
5. For NE Monsoon, the highlands of Eastern Ghats play the role of orographic barrier.

Features of Monsoon Rains

There are some unique features about the rains that Monsoon brings to the Indian subcontinent.

"Bursting" of Monsoon Rains

Bursting of Monsoon implies the onset of the sudden change in weather conditions in India (typically from hot and dry weather to wet and humid weather during the SW Monsoon) due to abrupt rise in the *mean daily rainfall.*

Similarly the burst of NE monsoon marks an abrupt increase in the *mean daily rainfall* over the affected regions.

Monsoon Rain Variability ("Vagaries")

One of the most commonly used phrases to describe the erratic nature of the Monsoon of the Indian subcontinent is "vagaries of monsoon", used in newspapers, magazines, books, web-portals to insurance plans and India's budget discussions.

In some years, it rains too much causing floods in several parts of India, in others it rains too little or not at all causing droughts. In some years when the rain quantity is sufficient, its timing may be arbitrary.

In some years, in spite of *average annual rainfall,* its daily distribution or the areal distribution might be substantially skewed. Such is the variability in the nature of Monsoon rains and weather.

Ideal and Normal Monsoon Rains

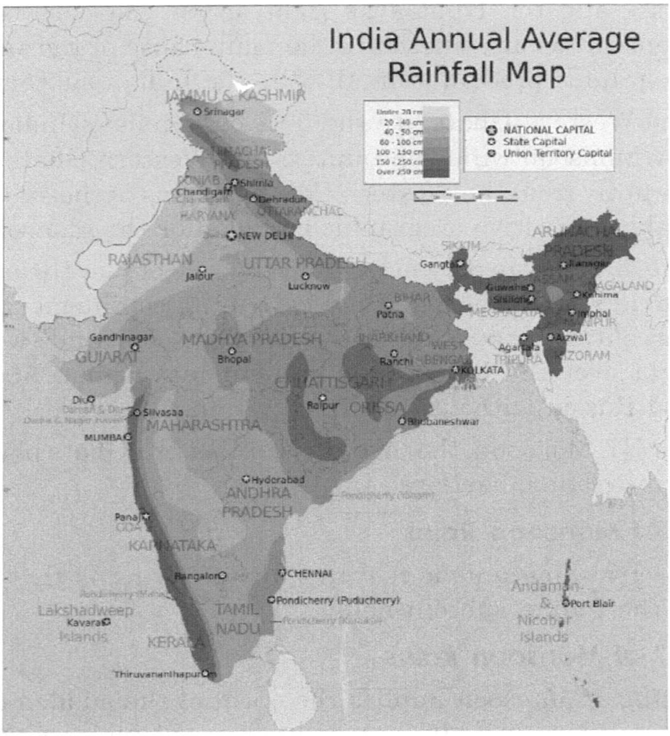

Figure: *Annual Average Rainfall Map of India*

Every year the normal onset of SW Monsoon is expected to *"burst"* onto the western coast of India (near Thiruvananthapuram) around 1 June covering entire India by around 15 July. Its withdrawal from India typically starts from 1 September onwards and completes by around 1 October.

Similarly the NE Monsoon is expected to *"burst"* around 20 October and last for a period of about 50 days before withdrawing.

However, a rainy Monsoon is not necessarily a *normal* Monsoon. A *normal Monsoon* is expected to perform close to its statistical averages calculated over a significantly long periods. Therefore, a *normal Monsoon* is generally accepted to be the Monsoon that has near *average quantity* of precipitation over all the geographical locations (*mean spatial distribution*) under its influence and over the entire expected time period of its influence (*mean temporal distribution*). Additionally, the *arrival date* and the *departure date* of both the SW and NE Monsoon should be close to the *mean* dates. The exact criteria for *Normal Monsoon* is defined by the Indian Meteorological Department with calculations for the *mean* and *standard deviation* for each of the aforesaid precipitation variables.

A Monsoon with excess rain can cause floods in India, Pakistan and Bangladesh. and one with too little rain can lead to widespread drought, food shortage, famine and economic losses. Therefore, a *normal Monsoon* with *mean performance* is the most desirable Monsoon.

Theories for Mechanism of Monsoon

Theories for mechanism of Monsoon primarily try to explain the reasons for the seasonal reversal of winds and the timings of their reversal.

Traditional Theory

Due to difference in the specific heat capacity of land and water, continents heat up faster than the seas. Consequently the air above the coastal lands heats up faster than air above seas. This creates areas of low air pressure above coastal lands compared to the air pressure over the seas causing winds to flow from the seas onto the neighbouring lands. This is known as sea breeze

Process of Monsoon Creation

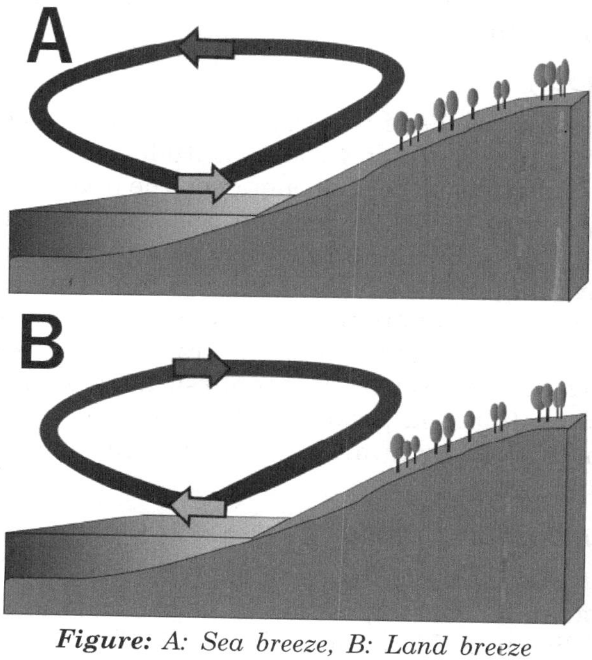

Figure: *A: Sea breeze, B: Land breeze*

Also known as the *thermal theory* or the *Differential Heating of Sea and Land Theory*, it portrays the Monsoon as a large-scale sea breeze. It states that during the hot sub-tropical summers, the massive landmass of Indian Peninsula heats up at a different rate than the surrounding seas resulting in a pressure gradient from South to North. This causes flow of moisture laden winds from sea to land. On reaching the land these winds rise up due to the geographical relief, cooling adiabatically and leading to orographic rains. This is the *southwest monsoon*. Reverse happens during winter when the landmass is colder than the sea establishing a pressure gradient from land to sea. This causes the winds to blow over Indian landmass towards Indian Ocean in a north-easterly direction causing the *northeast monsoon*. Since the SW monsoon is from sea to land, it has more moisture (therefore causing more rain) than the NE monsoon. Only a part of the NE monsoon passing over Bay of Bengal picks up moisture causing rain in Andhra Pradesh and Tamil Nadu during the winter months.

However many meteorologists argue that the Monsoon is not a local phenomenon as explained by the traditional theory but a general weather phenomenon along the entire tropical zone of earth. This criticism, does not deny the role of differential heating of sea and land in generating monsoon winds but merely restricts it to one of the several factors rather than the only one.

Dynamic Theory

The prevailing winds of the atmospheric circulation arise due to the difference in pressure at various latitudes of Earth and act as means for distribution of thermal energy on the planet. This pressure difference is due to the differences in Solar insolation received at different latitudes of Earth and the resulting uneven heating of the planet. Alternating belts of high-pressure and low-pressure develop along the equator, the two tropics, the Arctic and Antarctic circles and the two polar regions giving rise to the Trade winds, Westerlies and the Polar easterlies. However, the geophysical factors like revolution of earth, its rotation and axial tilt of the Earth result in gradual shifting of these belts northwards and southwards following the Sun's seasonal shifts.

Process of Monsoon Creation

The *dynamic theory of Monsoon* explains monsoon on the basis of the annual shifts in the position of global belts of pressure and winds. According to it, Monsoon is the result of the shift of the Inter

Tropical Convergence Zone (ITCZ) under the influence of the vertical sun. Though the mean position of the ITCZ is taken as the equator it keeps shifting northwards and southwards with the migration of the vertical sun towards the tropics (Tropic of Cancer and Tropic of Capricorn) during the summer of the respective hemispheres (Northern and Southern Hemisphere). As such, the theory states that during the northern Summer (months of May and June), the ITCZ moves northwards, along with the vertical sun, towards the Tropic of Cancer. The ITCZ being the zone of lowest pressure in the tropical region, is the target destination for the Trade winds of both the hemispheres. Consequentially, with ITCZ at the Tropic of cancer, the South East Trade winds of the Southern Hemisphere have to cross the equator to reach the ITCZ. However, due to Coriolis effect, (Coriolis effect causes winds in northern hemisphere to turn to its right whereas winds of southern hemisphere to turn to its left) these South East trade winds are deflected eastwards in the Northern Hemisphere transforming into South West trades. These pick up the moisture while travelling from sea to land and cause orographic rain once they hit the highlands of the Indian Peninsula. This results in the South-West Monsoon.

The *dynamic theory* provides the explanation of the system of Monsoon as a circum-global weather phenomenon rather than just a local one. And when coupled with the *Traditional Theory* (based on heating of Sea and Land) it enhances the explanation of the differential intensity of precipitation impact of Monsoon along the coastal regions with orographic barriers.

Jet Stream Theory

This theory tries to explain the establishment of both the NE and SW Monsoons as well their unique features like *bursting* and variability. The jet streams are a system of upper-air westerlies. It gives rise to slowly moving upper-air waves, with 250 knots winds in some air streams. First observed by World War II pilots, they develop just below the tropopause over areas of steep pressure gradient on the surface. The main types are the *polar jets*, the *subtropical westerly jets* and the less common *tropical easterly jets*. They follow the principle of geostrophic winds.

Process of Monsoon Creation

Over India, a *subtropical westerly jet* develops in the winter season which is replaced by the *tropical easterly jet* in the summer season. The high temperature over the Tibetan Plateau, as well as over Central Asia in general, during the summer is believed to be the critical factor leading to the formation of the *tropical easterly jet* over India in summer. The mechanism affecting monsoon is that the *westerly jet* causes high pressure over northern parts of the subcontinent during the winter. This results in the north to south flow of the winds in the form of the NE Monsoon. With the northwards shift of the vertical sun, this jet shifts northwards too. The intense heat over the Tibetan Plateau, coupled with associated terrain features of high altitude of the plateau, etc. generate the *tropical easterly jet* over central India. This jet creates a low pressure zone over the northern Indian plains influencing the wind flow towards these plains, assisting the establishment of the SW Monsoon.

Theory for "Bursting" of Monsoon

The unique feature of *bursting* of the Monsoon is primarily explained by the *Jet Stream theory* and the *Dynamic Theory*.

Dynamic Theory

According to this theory, during the summer months of Northern Hemisphere, the ITCZ shifts northwards pulling the SW Monsoon winds onto the land from the sea. However the huge landmass of the Himalayas continue to restrict the low pressure zone onto the Himalayas itself. It is only when the Tibetan Plateau heats up a lot more than the Himalayas does the ITCZ abruptly and swiftly shift northwards leading to *burst of Monsoon showers* over the Indian subcontinent. The reverse shift takes place for the NE Monsoon winds leading to a second minor *burst* during the Northern Hemisphere winter Months of NE Monsoon rainfall over Eastern Indian peninsula.

Jet Stream Theory

According to the theory the onset of SW Monsoon over Indian subcontinent is driven by the shift of the *subtropical westerly jet* northwards from over the plains of India towards the Tibetan Plateau. This shift is due to the intense heating of the Plateau during the summer months. This shift of the *westerly jet* to the north of the Himalayas is not a slow and gradual process, as expected for most changes in weather pattern. The primary cause of these is believed to be the height of the Himalayas. As the Tibetan Plateau heats up the low pressure created over it pulls the *westerly jet* northwards. Due to lofty Himalayas, the *westerly jet* is inhibited from moving northwards. However, with continuous dropping pressure, sufficient force is created for the movement of the *westerly jet* across the Himalayas after a significant period. As such, the shift of the jet is sudden and abrupt causing the *bursting of SW Monsoon rains* onto the Indian plains. The reverse shift happens for the NE Monsoon.

Theories for Monsoon Variability

The Jet Stream Effect: The above mentioned Jet Stream theory also explains the variability in *timing* and *strength* of the Monsoon.

Timing: A timely northward shift of the *subtropical westerly jet* in the beginning of the summer season is critical to the onset of the SW Monsoon over India. If the northward shift of this jet is delayed, so is the SW Monsoon. An early shift heralds in an early Monsoon.

Strength: Additionally, the strength of the SW Monsoon is determined by the strength of the *easterly tropical jet* over central India. A strong *easterly tropical jet* results in a strong SW Monsoon over central India and vice-versa.

El Niño-Southern Oscillation (ENSO) Effect

El Niño is a 'warm' ocean current originating along the coast of Peru that replaces the usual 'cold' Peru or Humboldt Current. This warm surface water reaching towards the coast of Peru with El Niño is pushed westwards by the trade winds thereby raising the temperature of the southern Pacific Ocean. A reverse condition is known as La Niña.

Southern Oscillation, a phenomenon first observed by Sir Gilbert Thomas Walker Director-General of Observatories in India, refers to the seesaw relationship of atmospheric pressures between Tahiti and Darwin, Australia. He noticed that when it was high pressure in Tahiti, it was low pressure in Darwin and vice versa. A *Southern Oscillation Index* (SOI),based on the pressure difference between Tahiti and Darwin, has been formulated by the Bureau of Meteorology (Australia) to measure the strength of the Oscillation. Walker noticed that the quantity of rainfall in the Indian subcontinent was often negligible in the years of high pressure at Darwin (and low pressure at Tahiti). Conversely, low pressure at Darwin bode well for the precipitation quantity in India. Thus he established the relationship of Southern Oscillation with quantities of Monsoon rains in India.

Ultimately, it was realized that the Southern Oscillation is just the corresponding atmospheric component of the El Niño/La Niña effect (which happens in the Ocean). Therefore in the context of the Monsoon, the two cumulatively came to be known as the *ENSO*. The ENSO is known to have a pronounced effect on the strength of SW

Monsoon over India with the Monsoon being weak (causing droughts in India) during the El Niño years whereas La Niña years had particularly good Monsoon strength over India.

Indian Ocean Dipole Effect

Although ENSO was statistically effective in explaining several past droughts in India, in the recent decades the ENSO-Monsoon relationship seemed to weaken in the Indian subcontinent. For e.g. the 1997, strong ENSO failed to cause drought in India. However, it was later discovered that just like ENSO was an event in he Pacific Ocean, a similar seesaw ocean-atmosphere system in the Indian Ocean was also at play. It was discovered in 1999 and named the Indian Ocean Dipole (IOD).

An index to calculate it was also formulated. IOD develops in the equatorial region of Indian Ocean from April to May peaking in October. With a positive IOD winds over the Indian Ocean blow from east to west. This results in the Arabian Sea (western Indian Ocean near African Coast) being much warmer and eastern Indian Ocean around Indonesia becoming colder and dry.

In the negative dipole year, reverse happens making Indonesia much warmer and rainier. It was demonstrated that a positive IOD index often negated the effect of ENSO, resulting in increased Monsoon rains in several ENSO years like the 1983, 1994 and 1997. Further, it was shown that the two poles of the IOD - the eastern pole (around Indonesia) and the western pole (off the African coast) were independently and cumulatively affecting the quantity of rains for the Monsoon in the Indian subcontinent.

Equatorial Indian Ocean Oscillation

Similar to ENSO, the atmospheric component of the IOD was later discovered and named as *Equatorial Indian Ocean Oscillation (EQUINOO)*. When EQUINOO effects were factored into the statistics certain failed forecasts, like the acute drought of year 2002, could be further accounted for. The relationship between extremes of the Indian summer monsoon rainfall, along with ENSO and EQUINOO have been studied and models for enhanced predictability of the quantity of monsoon rains have been statistically derived.

Monsoon Rain Prediction Models

Since the Great Famine of 1876–78 in India, various attempts have been made to predict the rainfall during the Monsoons in India. At least five models for prediction of Monsoon rains exist in India.

Seasonal Prediction of Indian Monsoon (SPIM)

Centre for Development of Advanced Computing (CDAC) at Bengaluru is facilitating the Seasonal Prediction of Indian Monsoon (SPIM) experiment on the PARAM Padma supercomputing system. This project did simulated runs on historical data from 1985 to 2004 to try and establish the relationship between of five *atmospheric general circulation models* with the Monsoon rainfall distribution.

Indian Meteorological Department Model

IMD has tried to forecast the Monsoon for India since 1884, some unsuccessfully but till 2011 is the only official agency entrusted with making public forecasts about the quantity, distribution and timings of the Monsoon in India.

IMD's position as the sole authority on the matter was further reiterated in 2005 by the Department of Science and Technology (DST), New Delhi. In 2003, IMD underwent a substantial change in its forecast methodology, model as well as its administration. A sixteen parameter monsoon forecasting model used by the Indian met office since 1988 was revamped in 2003 with a new one. However, following the 2009 drought in India (worst since 1972), IMD decided in 2010 that it needed to develop an *indigenous model* to further enhance its prediction capabilities.

Significance

Indian Monsoon is the primary delivery mechanism for fresh water in the Indian subcontinent. As such it impacts the environment (and associated flora, fauna and ecosystems), agriculture, society, hydro-power production and geography of the subcontinent (like availability of fresh water in water bodies, underground water table) with all these factors cumulatively contributing towards the health of the economy of affected countries.

The Indian Monsoon turns large parts of India from a kind of semi-desert into green grasslands.

Geographical (Wettest Spots on Earth)

Mawsynram and Cherrapunji, both in the Indian state of Meghalaya alternate to be the *wettest places on Earth* given the quantity of their rainfall. There are other cities with similar claims but with 10,000mm of rain for each of these locations, Monsoon of Indian subcontinent is a significant contributor towards the water supply of its area of influence.

Agricultural

India, historically an agrarian economy primarily, has recently seen the services sector overtaking the farm sector in terms of GDP contribution. However, even today agriculture sector contributes 17-20% of GDP and is the largest employer in the country with about 60% of people dependent on it for employment and livelihood. The land use pattern of India indicates that 49% of land is under agriculture in India, it is 55% if associated wetlands agriculture, dryland farming areas, etc. are included.

Since over half of these farmlands are rain-fed, Monsoon is critical to the food sufficiency and quality of life for the country. Despite progress in alternative forms of irrigation, agricultural dependency on monsoon is far from insignificant, even today.

Therefore, the agricultural calendar of India is governed by Monsoon. Any fluctuations in the time distribution, spatial distribution or quantity of the monsoon rains may lead to conditions of floods or droughts causing the agricultural sector to adversely suffer. This has a cascading effect on the secondary economic sectors, the overall economy, food inflation and therefore the overall quality and cost of living for the general population in India.

Economic

The economic significance of monsoon can be aptly summed up by Pranab Mukherjee's statement that monsoon is the *real finance minister of India*. A good monsoon resulting in improved agricultural brings down prices of essential food commodities and reduces their imports overall reducing the food inflation. Further improved rains result in increased hydroelectric production. All these factors initiate positive ripple effects throughout the economy of India.

Social

D. Subbarao (Governor of Reserve Bank of India), during a quarterly review of the monetary policy, once highlighted that lives of Indians depends on the performance of Monsoon. His own personal career prospects, emotional well being and the performance of his monetary policy were all *a hostage to monsoon* like it was for most Indians. Additionally, farmers, rendered jobless due to failed Monsoon rains tend to migrate city-wards. This crowds the city slums and further aggravates the job, infrastructure and sustainability of city life. Such is the magnitude of effect that monsoon casts on the lives of Indians.

Travel

In past, people usually refrained from travelling during monsoons for practical as well as religious reasons. But with advent of globalization, travel during monsoons is gaining popularity. Places like Kerala, Western Ghats get a very large number of tourists, both local and foreigner during monsoon season. Kerala is one of the top destinations for tourists interested in Ayurvedic treatments and massage therapy. One major drawback of travelling during monsoons is the fact that most of wildlife sanctuaries are closed during this rainy season. Also, some mountainous areas, specially in Himalayan regions get cut off due to damaged roads caused by landslides and floods during heavy rains.

Environmental

The Monsoon is the primary bearer of fresh water to bodies of water in the area. The peninsular/Deccan rivers of India are mostly rain-fed and non-perennial in nature depending primarily on the Monsoon for water supply. Similarly most of the coastal rivers of Western India are rain-fed and Monsoon dependent. As such, obviously the flora, fauna and the entire ecosystem of these areas are heavily dependent on the Monsoon.

Dryland Agriculture and Conservation Technologies for Improving Soil Moisture for Crop Production in Semiarid Kenya

Dryland agriculture is practised where there is water stress and seasonal agricultural drought due to low water supply (low rainfall, high runoff water losses and high evaporation). In dryland agriculture, emphasis is on conservation of soil water for crop production. Dryland agriculture is defined as land husbandry under conditions of moderate to severe soil moisture stress during a substantial portion of the year, which requires special cultural techniques, adapted crops and sustainable production systems. Typical characteristics of dryland areas include: low total rainfall with at least one pronounced dry season; highly variable and unreliable rainfall during the normal rainy season with temporal (year to year, month to month) variations in rainfall amount and distribution; increase in persistence of agricultural drought with decreasing rainfall; potential evapo-transpiration exceeding rainfall for at least seven months of the year; and very high intensity rainstorms of short duration which often lead to high surface runoff and soil erosion. Thus limited soil water supply and a serious erosion potential are the two most common characteristics in all dryland areas.

The three broad areas of soil, water and nutrient management in drylands include: management of soil water; management of soil erosion; and the management of fertilization. Less soil water could lead to some water stress. Likewise less fertilization could lead to some nutrient stress. Severe soil erosion would result in the loss of nutrient rich topsoil.

In dryland areas, the most dominant soil types - Lixisols, Luvisols, and Acrisols (FAO/UNESCO Classification) are of limited agricultural productivity due to soil related problems. Thus the four major problems limiting dryland crop production are severe soil erosion, low soil fertility, low soil moisture and high soil crusting and compaction. These four soil management problems are stratified according to slope, agro-climatic zones, land use and soil type. The occurrence of soil crusting and compaction is attributed to seasonal rainfall characteristics (amount, intensity and duration), physical soil properties and bad tillage practices. The problem of soil compaction is caused by bad tillage practices (leading to the development of plough pans) and natural soil conditions [subsurface hardpans due to clay eluviation (leaching) from the A to B horizons]. Soil crusting and compaction decreases rainwater infiltration and increases surface runoff. Increased and concentrated surface runoff water flow causes severe soil erosion. Of the four soil related problems, Low soil moisture is the most critical as it directly affects dryland agriculture.

The effectiveness of soil, water and nutrient management interventions in dryland areas could be objectively assessed in terms of: enhanced infiltration of rainwater; increased soil water storage capacity; improvement of effective use of stored water; minimization of evaporation from the soil; and minimization of soil loss and salinization.

In order to realize the above objectives and enhance soil and crop productivity, there is need to increase rainfall and crop water use efficiency (RWUE and CWUE) and fertilizer use efficiency (FUE). These are verifiable indicators of effective soil and water management in dryland areas. Rainfall water use efficiency could be improved through: runoff waterharvesting and conservation (RWHC) for crop and fodder production; crop rotation where deep rooted plants would use water in deeper soil horizons; surface residue mulching and cover cropping to decrease surface evaporation; enhancement of infiltration and reduction in surface runoff; and ultimately an increase in crop water use efficiency. An increase in crop water use efficiency could be realized through: enhanced transpiration; application of fertilizer and organic manure; and enhancement of drought tolerance by crops. Likewise a high fertilizer use efficiency when there is some good crop response due to availability of nitrogen, phosphorus and potassium.

In semi arid Kenya, rainfall is bimodal, and characterized as low, erratic and poorly distributed. The short and long rainy seasons receive about 55% and 45% of the total annual rainfall respectively. The short rains (October to December) are more reliable, evenly distributed and adequate for crop production. The long rains (March to May) are associated with most crop failures due to the poor distribution, unreliability and inadequacy for crop production. The dominant soil types e.g. Lixisols, Luvisols, and Acrisols (FAO/UNESCO Classification) differ widely in depth, texture, chemical fertility, organic matter content and physical properties. In this area, soil crusting and compaction is a common soil management problem and is attributed to structural instability of the dominant soils (Luvisols, Lixisols, and Acrisols) and bad tillage practices. The problem of soil compaction in Iiuni is caused by bad tillage practices (leading to the development of plough pans) and natural soil conditions [subsurface hardpans due to clay elluviation (leaching) from the A to B horizons].

So what are the on-farm conservation tillage technologies that have been tested under dryland conditions. The tillage methods that have been tried include zero tillage, conventional tillage with two farmyard manure (5 and 10 Mg/ha). Farmyard manure (FYM) when incorporated into the soil enhanced infiltration and reduced soil crusting and compaction, and surface runoff during the initial stages of a rainy season. In the mid-stages of the rainy season, the effects of soil aggregation from FYM diminished and this resulted in increased soil loss. Towards the end of the rainy season when soil crusts had formed, some significant decrease in soil loss with FYM treatments was observed. Conventional tillage without farmyard manure led to high surface runoff and soil loss for Luvisols (with inherent low aggregate stability). Zero tillage has performed poorly under these soil conditions because of the significantly high soil crusting and compaction, surface runoff and subsequent low rainwater infiltration.

Recently (1999-2001) three tillage techniques that included conventional tillage, subsoiling and ripping/ridging and pitting/ridging were tried out through on farm demonstration trials in a semi arid area of Kenya (Biamah, 2001). Conventional tillage was done using the conventional ox-drawn mouldboard plough to a maximum depth of 10 cm. Below this depth, is a plough pan that often developed due to compaction from continuous conventional tillage operations.

Subsoiling and ripping/ridging was done on fallow land using an animal drawn subsoiler and a ripper/ridger.

The subsoiling was done in strips along planting rows to a maximum depth of 25 cm. The ripper/ridger was used to increase the width of cut to about 5-6 cm up to the tillage depth of 25 cm and this created plain ridges for increased depression storage of surface runoff. Pitting and ridging was done on fallow land.

The pits were manually dug using hand hoes. Each pit was dug to a diameter of 10 cm and depth of 20 cm. The labour involved was quite intensive and costly.

In each pit, 2 to 3 handfuls of farmyard manure (FYM) were applied and mixed with the topsoil before planting. Some depression was maintained in each pit to impound runoff water during the rainy period.

From the statistics obtained during the short rains (2000/2001) and the long rains (2001), it was evident that pest (white grub and termite) damage accounted for about 18% to 28% of the total crop losses in Iiuni, Machakos.

Table: *Plant population, crop damages and percent losses due to termite activity during the short rainy season, 2000-2001 at Iiuni, Machakos, Kenya (Biamah, 2001).*

CONTIL technique	***Plant population (No)***	***Crop damage (No)***	***Crop loses (%)***
Conventional tillage	990	275	28
Animal subsoiling and ripping	965	265	27

Table: *Plant population, crop damages and percent losses due to white grub and termite activity during the long rainy season, 2001 at Iiuni, Machakos, Kenya (Biamah, 2001)*

CONTIL technique	***Plant population (No)***	***Crop damage (No)***	***Crop loses (%)***
Conventional tillage	714	141	20
Animal subsoiling and ripping/ridging	772	128	17

- The crop yields presented in Tables, show a trend of good grain and biomass yields during the short rainy seasons of 1999 and 2000. Under these semi arid rainfall conditions, these yields are above normal.
- The maize grain yields obtained by conservation tillage farmers during the short rainy season of 2000 were the highest in several decades of their farming career.

- The crop yields for the long rainy seasons (2000 and 2001) were good when compared with what a conventional farmer would get. During the two long rainy seasons, no conventional farmer harvested any maize grain.

Table: *Average grain and biomass yields during the short rainy season, 1999-2000 at Iiuni, Machakos, Kenya (Biamah, 2001).*

CONTIL technique	Grain yield (kg/ha)	Biomass yield (kg/ha)
Conventional tillage	1700	2560
Animal subsoiling and ripping	2580	3080

Table: *Average grain and biomass yields during the short rainy season, 2000-2001 at Iiuni, Machakos, Kenya. (Biamah, 2001).*

CONTIL technique	Grain yield (kg/ha)	Biomass yield (kg/ha)
Conventional tillage	2893	3268
Animal subsoiling and ripping	3121	3290

Table: *Average grain and biomass yields during the long rainy season, 2000 at Iiuni, Machakos, Kenya. (Biamah, 2001).*

CONTIL technique	Grain yield (kg/ha)	Biomass yield (kg/ha)
Pitting and ridging	2875	2300

Table: *Average grain and biomass yields at Iiuni, Machakos, Kenya. Long rainy season, 2001 (Biamah, 2001).*

CONTIL technique	Grain yield (kg/ha)	Biomass yield (kg/ha)
Conventional tillage	342	1268
Animal subsoiling and ripping/ridging	385	1529

Farmers Perceptions

Farmer experiences with conventional tillage practices in the trials area showed that improvements in soil and crop productivity occur only when the soil erosion problem is minimized first through effective conservation structures. Thus conservation tillage practices are complimentary to erosion control and provide remedies to soil fertility, soil moisture and soil compaction problems. Thus conservation tillage is viewed as a viable land management option for improving soil moisture conservation for dryland crop production in semi arid Kenya.

Conclusions

- In the conservation tillage trials area, conventional tillage operations using the ox-drawn mouldboard plough resulted in the development of plough pans below a maximum ploughing depth of 10 cm. Besides there was the natural subsurface hard pan that is attributed to clay eluviation in the B_t horizon.
- The breaking of subsurface hardpans calls for conservation tillage techniques that improve the depth of tillage to 25~30 cm and thus enhancing the soil moisture storage in the soil. The depth of tillage in the trials area varied with the tillage implements used and soil conditions especially the antecedent soil moisture conditions.
- Deep or shallow tillage could be recommended on the basis of seasonal soil conditions as influenced by rainfall characteristics. For instance, deep and shallow tillage are recommended for the short and long rainy seasons respectively. Deep tillage is an effective tillage operation for enhancing soil moisture storage during the long dry spell before planting. Likewise shallow tillage is an appropriate tillage operation for minimizing soil moisture losses through evaporation.

Dryland Cropping Systems

Quick Facts...

- Soils under no-till production systems store more water than soils on conventional stubble mulch systems and allow conversion to more intense crop rotations.
- Conversion to no-till systems using a three- to four-year rotation such as wheat-corn-fallow, wheat-millet-fallow, or wheat-corn-millet-fallow, results in more profitability.
- Conversion to no-till, three- to four-year rotations results in greater risk and requires greater management input.

The Wheat-Fallow (WF) system, practiced for many years in the semiarid Western Great Plains, was a definite improvement over continuous cropping. It stabilized yields and provided farmers with a reliable income from year to year. However, wheat fallow is an inefficient user of annual precipitation in a region where water is the major limiting factor. Data collected in the Great Plains show that typically less than 25 percent of the precipitation received during the 14-month fallow period is stored in the soil. When little or no residue is left on the soil surface, water storage efficiency is less than 20 percent.

Colorado and Nebraska research indicates that precipitation storage efficiencies of 40 to 60 percent is achieved when tillage is minimized or eliminated. The key to these improvements is maintaining crop residue on the soil surface and minimizing or eliminating soil disturbance. Unfortunately, little added wheat yield results from the additional water stored under no-till compared to mulch tillage systems. In fact, it costs more to save the additional water in a no-till system than the value of the added grain yield. Weed control with herbicides is more costly than with mechanical tillage.

Water savings in no-till systems can be converted to profit only by switching to more intense cropping systems where fallow time is decreased and summer Crops, like corn, grain sorghum, proso millet and annual forages are added to the rotation.

Maximizing water storage, which permits more intensive rotations, requires residue maintenance on the soil surface to trap snow, absorb raindrop impact, slow runoff and minimize evaporation, as well as complete weed control at critical times in the production system.

Producers can use the following recommendations to improve precipitation use efficiency in eastern Colorado dryland farming systems, and improve soil, water and air quality while increasing net income. These steps are built around rotations where winter wheat is the starting point in the system.

How to Intensify Dryland Crop Rotations

Weed Control: Weed control in growing wheat is the goal of this step. If controlled in the wheat, the stubble at harvest is essentially weed free. Weeds include winter annuals such as mustard, kochia and Russian thistle.

If weeds are present at harvest, apply the appropriate herbicide for that species. Apply residual herbicides in late July to mid-August to control volunteer wheat, downy brome and jointed goatgrass. Various combinations of herbicides with small amounts of atrazine are effective for this purpose.

Keep the atrazine rates low and within labelled amounts for your soils and climate. Failure results in long-term damage to wheat yields.

Do not till in this phase of the rotation. Dollars spent for herbicides in this phase give maximum returns compared to any other point in the system. It should be possible to store 3 to 5 inches of water between wheat harvest to fall freeze-up if the stubble is kept weed free. By not tilling, the maximum amount of standing stubble can trap snow in the winter months and increase water storage by spring.

Spring Crop Choices and Cultural Practices: Possible summer Crops to follow wheat in the rotation are corn, sorghum, proso millet, sudex or sunflowers. Plant corn in areas north of Cheyenne Wells and grain sorghum in southern areas. Proso millet is a good option in the northern area, but not in the southern areas where yields and markets are poor. Annual forages, like sudex and hay millet, are excellent choices for livestock producers.

Whatever the summer crop, plant it into the wheat stubble or other residue base with minimal disturbance. A minimum till or slot planter is recommended. Apply an appropriate herbicide according to the label and use for weed control during the summer period. Avoid cultivation of row Crops unless the herbicide fails. Choose crop varieties that adapt to your elevation, latitude and longitude.

Rotation Choice: Specific rotation choices include: wheat-corn-fallow; wheat-sorghum-fallow; wheat-proso millet-fallow; wheat-corn-proso millet-fallow; etc. More intensive rotations (wheat-corn-proso millet-forage) are possible where no fallow is used. Consider any combination of Crops that adapt to your environment and fit your equipment, budget, and overall livestock feed requirements. Do not follow a particular crop with itself. There are too many insect, disease and volunteer problems when you repeat a species in a rotation.

Research suggests that corn follow wheat and proso millet, or an annual forage follow corn or grain sorghum. If you intend to keep winter wheat in your rotation, some fallow prior to planting winter wheat probably will be necessary. Another way to get back to winter wheat is to plant a short-season annual forage in the spring and harvest the forage prior to August 1. If regrowth occurs, kill immediately with a contact (no residual) herbicide so that all precipitation received after August 1 is stored for wheat planting. Wheat should be planted no-till into the forage stubble. If it was a dry summer and no stored soil moisture is present, you may have to wait until spring and go back to another summer crop.

Fertilization: Nitrogen, phosphorus and zinc are the nutrients of concern when making fertilizer decisions. Zinc deficiency on wheat has never been observed in Colorado, but it is common for corn and sorghum. Wheat is the crop most likely to respond to phosphorus fertilization.

Base the amount of fertilizer to apply, if any, on soil test results. Ideally, collect separate samples from various soils on your farm. Sample eroded hillsides separately from summits and separately from lower-lying areas. Soil colour is a good way to judge whether soils should be sampled separately.

Apply fertilizer according to each soil test if possible. This will result in the least-cost fertilizer programme.

Overall Considerations: All of the suggestions for intensifying cropping under dryland conditions are contingent upon maximum water capture and minimum losses to weeds, runoff and evaporation. No-till practices are required in certain phases of each system to be reasonably sure of conserving enough water to be successful. No-till is essential, for example, when going from wheat to corn or from corn to proso millet. It is feasible and most profitable, however, to substitute some tillage for herbicides in a summer fallow phase preceding wheat planting. An exception is a short fallow period from harvest ofa summer forage crop in August to a September wheat planting. No-till is the only feasible means of weed control in that situation.

3

Development of Drylands for Conservation of Biodiversity and Food Security

Drylands in India contribute to over 70% of the total cultivated area and about 50% of the total geographic area is affected by desertification. Drylands are not static. These are subject to various pressures, which lead to further deterioration. Hence, drylands when neglected, may slowly turn into deserts. Desertification is the result of many factors such as low rainfall, poor water conservation measures, heavy soil erosion and harsh climatic conditions. Biotic pressure further accelerates the process of desertification. Unfortunately, most of the arable drylands are owned by poor farmers who are unable to earn sustainable livelihood from such lands. Major crops grown on drylands are minor millets, pulses and oilseeds, which are drought tolerant and withstand moisture stress, nutritional deficiencies and biotic pressure. Crop production on drylands is risky and the chances of failure are almost 2-3 times in a block of 5 years. Therefore, in the absence of assured income from agriculture, most of the families living in dry regions maintain different types of livestock for supplementary income.

Apart from arable drylands, there are wide stretches of community pastures, village common lands and degraded forest lands, falling under the category of drylands. Out of the total wastelands spread over 135 million ha, at least 50% of the area is suffering from moisture stress and unable to support plant growth. Thus, these wastelands owned by public institutions, also fall under the category of drylands. As the drylands are closely associated with poverty, food security and backwardness, neglect of drylands will not only turn these valuable assets into deserts but also invite disaster for our ancient civilized society. Therefore, development of drylands deserves priority.

Causes of Dryland Formation

Lack of moisture to support the growth of plants and micro-organisms is the primary cause of dryland formation. Areas receiving less than 300 mm of rainfall per annum, face severe moisture scarcity during a major part of the year. It is very difficult to take up profitable agriculture on such lands without supplementary irrigation. There are many such regions where shrubs, herbs and grasses grow profusely and dry out in autumn. Sahelian region of Africa located on the edge of Sahara desert, receives only 250-300 mm rainfall and provides livelihood to millions of families through livestock husbandry. Even in Western India, the families living in Thar region of Rajasthan and Banni region of Gujarat are dependent on animal husbandry for their livelihood, only because of good growth of seasonal grasses and herbs. Such lands may not support perennial shrubs due to severe moisture stress in summer.

Arid lands, receiving an annual rainfall of 300-500 mm can easily support one crop. Such areas are found in the states of Haryana, Rajasthan, Gujarat, Maharashtra, Andhra Pradesh and Karnataka. Some of these areas are now receiving irrigation and yielding 6-8 times more. With irrigation the employment potentials will also increase by 3-4 folds. The remaining dry areas still experience frequent crop failures, forcing the farmers to live in chronic poverty. Such poor farmers neither have resources to invest in critical agricultural inputs nor are they interested in taking any risk due to high chances of failure. Thus, the crop yields remain suppressed even in good years. It is a vicious cycle. Therefore, it is necessary to develop a sustainable model by introducing a mixed farming system with high stake in livestock husbandry to reduce the risk of failures. In such a system, even the failed crop residues turn out to be a boon for feeding livestock.

Apart from the lands suffering from moisture stress in arid regions, there are several other types of drylands located in different parts of the country. These are:

- Deep sandy soils with poor water holding capacity;
- Shallow and rocky soils which cannot absorb rain water;
- Deep soils with undulating surface, unable to hold rain water;
- Cold regions where plants cannot absorb water due to severe cold

- Sodic and saline wastelands, where plants cannot survive due to high salt content.

Problems of Drylands Development

Among various types of drylands, it is only the arid regions where the productivity is affected by moisture stress. The other types of drylands are remaining idle due to lack of our interest and attitude and posing a threat to our livelihood, eco-system and biodiversity. Lack of technologies, limitation of resources and biotic pressure contribute further in conversion of these drylands into deserts.

Land has tremendous potential to support plant life. In India, there is hardly any location where the rainfall is less than 300 mm. Hence, there is no excuse for letting over 100 million ha drylands remain idle. This reflects on the lack of interest to manage these precious resources.

Fortunately, we have good success stories from India and abroad, where traditional land use practices have been providing sustainable livelihood to the local communities. Looking to these experiences, we need to develop a strategy for developing the drylands for generating gainful employment to our rural poor, while enriching the environment.

In semi-arid regions where the rainfall is over 600 mm, retention of rainwater in the field is the major problem. Increased infiltration of rainwater, assures good moisture to the soil to support plant growth. Thus, water is the lifeline not only for the survival of human civilisation but also for keeping the land resources in good health. Unfortunately, water is becoming a scarce commodity. We have already been witnessing inter-state disputes for sharing river water. Thousands of villages are dependent on tankers for water supply in summer. Rural women still spend several hours to fetch water every day. Drinking water is costlier than milk in cities. It has been predicted that in future, the world wars will be fought for water. This reflects a grave situation.

However, in India, we have been witnessing the negative impact of water, where water is a curse for many. Over 8 crore people are affected by floods every year in the country. Last year had the dubious record of many mega cities like Mumbai, Chennai, Baroda and Kolkata sinking in floods. This is an indication of heavy soil erosion and runoff of rainwater in the catchment areas. Hence, floods are warning signals of dryland formation in the catchment areas. Our inability to harness the precious water resources should be a cause of serious concern for all those involved in community development.

Development of Drylands

Therefore, for the development of drylands, the primary step is to initiate a very efficient method of soil and moisture conservation. These include establishment of contour bunds, land shaping to retain rainwater, formation of farm ponds and percolation tanks. With efficient harnessing of rainwater, there will be a check on soil erosion, apart from recharging of the ground water table. The immediate benefit will be assured supply of potable water for the local population.

Farm ponds play a very effective role in recharging of groundwater, particularly in sandy and loamy soils. In Karnataka, BAIF has established a network of farm ponds covering a watershed of 3-4 villages. These farm ponds were inter-connected to ensure that the overflow of water from the upper level pond will go to the lower level pond through the channel. This network could prevent soil erosion to a great extent while facilitating the ponds to retain water for over 6-10 months in a year. As a result of effective recharging of ground water, the rivulets located on the lower evaluation in the network area could retain water throughout the year. This encouraged the farmers to establish plantation crops in the catchment area of the watershed, which was otherwise remaining idle. Another significant impact of the programme was revival of hundreds of borewells, which had dried out several years ago. This proved to be an excellent demonstration to motivate farmers to take up water conservation measures for their economic sustainability.

With the availability of water, drylands can be brought into various uses. At Urulikanchan, where the soil is very shallow, it was possible to support a wide range of crops. However, the economics of growing such crops is dependent on the quantity of water available to ensure optimum growth. In sandy desert areas of Kutchh, farmers have taken good initiative in establishing fruit crops like date palm and ber, using brackish water. Establishment of trees as windbreaks on the field bunds and borders further help in conservation of moisture by maintaining a low temperature. Thus, depending on the water availability and economic viability, a suitable agri-horti-forestry system can be established on drylands.

In the absence of any sources of water for irrigation, BAIF has promoted community pasture development in the semi-arid regions of Rajasthan. Under this programme, the primary step was to establish contour bunds to prevent runoff of rainwater. The other important activity was to dig the trench-cum-bund on the border and establish a live hedge fence to prevent biotic pressure. Subsequently, seeds of Dhaman grass (*Cenchrus setigirus*) and Stylos (*Stylosanthus hemata*) were sown after tilling the land lightly, using tractor-drawn tillers. This could help in revival of the pasture along with a wide range of native tree species. The local community could harvest 4-6 tons of grass per ha. The critical factors for success of this programme were people's participation, their willingness to control stray animals and share the resources equally among the members.

With assured sources of water, developed through watershed development, it is possible to establish good horticultural plantations in the areas, where the annual rainfall is above 800 mm. This programme will not only provide an excellent opportunity for employment generation for the rural poor, including the landless, but also help in conserving the eco-system and biodiversity. Based on this experience, BAIF has initiated the Tribal Development programme, popularly known as *Wadi*. Under this programme, drought tolerant fruit trees were selected as the main crops and the inter-space was used for cultivation of food crops, which not only ensured food security, but also provided gainful employment till the trees started bearing fruits. The borders were used for cultivating different types of multipurpose tree species useful for fodder, fuelwood, timber and herbal medicines. They also served as windbreaks and protection to the orchards. This programme, which enabled each target family to develop horticulture on 0.4 ha, was also able to provide a net income of Rs.20,000-25,000 per annum after 4-5 years. The programme is now recognised by the Ministry of Tribal Affairs, Government of India and NABARD for wider replication.

Role of Industries in Dryland Development

BAIF also had a unique experience of establishing industrial greenbelt on degraded lands in different locations. At Ghansoli village, in the Thane-Belapur industrial belt in New Mumbai, BAIF has reforested 150 ha degraded forest lands with the support of NOCIL. Although this area receives more than 2000 mm of rainfall, most of the hill ranges of the Western Ghats are heavily eroded and barren. Such lands are worse than drylands. These barren hills are causes of flash floods. Hence, staggered trenches were dug across the slope to facilitate percolation of rainwater. Looking to the water shortage during the summer, treated affluent from the petrochemical industry was used for irrigating the plantation. This gave a tremendous boost to the plantation and in three years time, the entire barren hill was converted into a popular spot for eco-tourism. Over 2000 people, from a radius of 50 km come to this spot for picnic on weekends. Our experience of working at Gokak Falls with the support of Gokak Mills, covering 150 ha was also very interesting. We could develop tree plantations without any irrigation. More interesting was the natural growth of neem trees, which have established into a thick grove over a period of 5-8 years.

BAIF has taken up similar greenbelt development work with the support of other industries, such as L&T at Hazira, IPCL at Nagothane and Bharuch, Hindustan Lever at Khamgaon, INDAL at Belgaum and NPCL at Kakrapara near Surat and Tarapur. These industrial greenbelts have created an excellent ambience in the surroundings and a very positive feeling among the staff working in the industry. The Managements of these establishments were able to appreciate the value for the money spent.

Reclamation of Degraded Lands

Ravine lands are also a type of drylands, where soil is sandy and the water retaining capacity is very low. These lands, being deprived of vegetation are prone to heavy erosion causing deep gullies. The major problem was the attitude of the local people, who did not consider it necessary to protect the community lands, adjoining their agricultural lands, which accelerated the formation of ravine lands. BAIF was able to gain confidence of the local community and demonstrated the need for restoration of community lands. Contour bunding and gully plugging could help in retaining rain water in tanks. As a result, the groundwater table could increase and the farming families were able to drill group borewells for irrigating their crops in two seasons. Today, these people are taking care of the ravine lands because they are aware that without protecting the common lands, they will not be able to harvest bumper crops on their agriculture lands.

Even the sodic and saline lands can be considered as drylands because they cannot nurture any greenery. The formation of sodic land is due to excessive use of water for irrigation without any provision for drainage. Unfortunately, we have 9 million ha of fertile lands with assured irrigation to grow 2-3 crops per year, which have turned into sodic lands. This indicates that we have wasted precious water, which otherwise could have brought additional 10 million ha under irrigation. This is an indication of the mismanagement of water. It is high time that the local communities are assigned the responsibility of sharing water and other natural resources equally and efficiently to ensure high productivity and social justice.

Call for Action

It is clear that most of the drylands in our country are our own creation. They can certainly be brought under productive use, which will not only conserve our eco-system and biodiversity and generate employment, but also ensure food security for millions of rural poor. This is not an exaggeration. In Israel, farmers and scientists have made excellent efforts to introduce many drought tolerant fruit trees in sandy deserts. In large parts of Europe, the hilly terrains are like drylands. Nevertheless, the farmers are making sustainable use of such hilly areas for livelihood through dairy husbandry.

Unfortunately, inspite of good rains and assured sources of water, we are not able to develop our drylands in many parts of the country. Local communities feel helpless because they lack technology, resources and measures to reduce the biotic pressure. It is due to our apathy and neglect that the drylands are in the process of desertification. We, therefore, need to initiate the development of these drylands through a mission mode with pilot projects, which can serve as models for wider replication.

If the Government of India is prepared to take initiative and permit agencies like BAIF to develop drylands including the degraded forest lands, and if Corporate houses come forward to sponsor such development programmes, we can certainly bring a change.

Soil Management for Dryland Area

Dry farming or dry land farming is the practice of crop production entirely with rain-water received during the crop season or on conserved soil moisture in low rainfall areas of arid and semi-arid climates and the crop may face mild to very severe moisture stress during their life cycle. Dry land farming is characterized by having:

a. limited rainfall less than 800 mm,

b. shortage of moisture availability,

c. growing season less than 200 days,

d. single crop or intercropping system and

e. having constrains of wind and water erosion.

Soil Types in Dryland Areas

Five types of soils observed are black, alluvial, red (laterite and lateritic), sierozems and submontane soils. In the north and northwest of the country, alluvial, sierozemic and submontane soils predominate.

In central and south India, black and red soils occupy the highest area. In the higher rainfall regions and coastal areas, the laterites and lateritic soils predominate.

Black Soils: The black soils are deeper, clay to clay loam and characterised by low permeability and high water holding capacity. Low infiltration rate, high plasticity and stickiness, low organic matter content, high CEC, the calcareous nature and slightly alkaline reaction, pose problems of management practices, Vertisoils, when kept fallow during Kharif, are exposed to soil erosion hazards.

Red Soils: The red soils are light textured, shallow to medium in depth and usually underlain by compact subsoil, fairly porous and low water holding capacity. Soils are prone to erosion and surface crusting. Because of crust formation run off Alfisols is more than in Vertisols. Crusting just after seeding results in the poor emergence of seedlings, particularly in the case of small seeded crops such as finger millet and pearl millet.

Alluviai Soils: These soils are fairly level, deep, light to medium in texture with favourable physical characteristics and good permeability. Small showers are useful and there is the utilisation of most of the water held by the soil due to low moisture content at wilting point.

Sierozemic Soils: Very deep alluvial sandy loams, low soil moisture storage, instability of soil structure, and poor soil fertility is the major problems of soil management in the desert ecosystem. High wind velocity leads to servere wind erosion. Soil drifting leads to soil and nutrient losses. Surface crust formation after sowing following light showers limits the desirable crop stand. These soils are observed in Dantiwada, Hisser and Jodhpur.

Submontane Soils: Such soils are distributed in the dry subhumid environment of Hosiarpur in Punjab and Rakh Dhiansar in Jammu and Kashmir and in the humid tract of Dehradun. The lands are sloping, the soils range from loamy sands to sandy loams, silty loams and clay loams with soil moisture storage capacity improving in that order. Soil crusting occur in soils of dry and sub-humid regions.

Constraints

Edaphic Problems associated with dryland farm.

a. Poor and marginal lands with soils low in fertility and productivity;
b. Uneven topography with high erodability;

c. Difficulty in workability particularly in Vertisols;
d. Shallow or very deep in depth with extreme permeability;
e. Low moisture storage and release capacity particularly in Alfisols;
f. Presence of dissolved injurious salts in ground water;
g. Problem soils with respect to soil reaction (pH) and high concentration of soluble salts in the surface soils;
h. Water logging in level lands; flooding and breaking small field bunds resulting in poor conservation of soil and water;
i. Movement of sand and soil;
j. High surface crusting that leads to poor crop stand and high cracking to a high rate of evaporation and mechanical injury to roots.

Management Techniques

Reduction of moisture loss due to Evaporation and Transpiration:

Following measures are taken to reduce the loss of moisture received by the soil.

a. growing early maturing adaptable crop varieties with a deep and ramified root system and with a reduced number, size and horizontally orientation of leaves.
b. maintaining optimum plant population per unit area.
c. sowing crops either in dry soil anticipating rainfall suitable for early crop establishment with the first shower, and subsequent growth and development with subsequent rainfall, or in optimum soil moisture but with a minimum expenditure of seasonal moisture for land preparation and sowing;
d. keeping the field free from weeds,
e. adoption mixed or intercropping to utilise the slow growth phase of wide spaced crops, to restore soil fertility and to check soil and water loss,
f. using mulches,
g. using agri-chemicals;e.g. anti-transpirants, plant modifiers or growth retardants, desiccants or defoliants, crop ripeners, anti-evaporants, antiseepage.

Cropping Systems

Cropping systems differ according to climate and soil types. The areas with 400 to 750 mm annual rainfall, mono cropping with traditional long duration crops is common. Generally adaptable crops are cercals, oil-seeds, pulses.

When the rainfall is between 500 to 700 mm with a distinct period of moisture surplus, the intercropping system can be adopted. Intercropping facilitates the growing of either cereal + legume or legume + legume. e.g. are: sorghum + pigeonpea, pearl millet + pigeonpea, sorghum + green gram, sorghum + soybean, groundnut + pigeonpea and foxtail millet + pigeonpea.

In areas with more than 750 mm annual rainfall with a soil storage capacity of 150 mm or more of available moisture sequential cropping is possible. e.g. pulses and oil-seeds, rice followed by chick-pea, maize followed by chick-pea, sorghum or green gram followed by safflower or sorghum- chick-pea and maize-chick-pea.

Mechanical Methods

- Advantages
  1. These break up the slope.
  2. Intercept runoff before its volume and velocity become sufficient to cause serious erosion
  3. gives more time for infiltration
  4. water is diverted into the channels down safe gradients of suitable discharge or outlet points which carry away water in such a way as to minimize erosion damage to other land and
  5. Finally leading to better conservation of run-off water for agriculture.
- Methods

Contour Bunding: The bund section is 1.61 m^2 in Vertisols and 1.05 m^2 in Alfisols. The vertical distance is about 0.9 m. The area occupied is upto 5.0 per cent by the bund and the area lost from cultivation due to stagnation in Vertisols would be 10 to 15 per cent.

Graded Bunding: Graded bunds are of 0.8 m^2 cross section in Vertisols and at vertical internal of 0.7m with a channel on the upstream side. The area lost due to the structure would be not more than 3-5 per cent and there would be no water stagnation and graded bund with grassed waterways and box-type masonry drainage outlets in arable fields.

Tie-ridging: The practice of tie-ridging, where adjacent ridges are joined at regular intervals by barriers or ties of the same height, allows the water to infiltrate and prevent run-off except during intense storms. This method is adequate in moderate rainfall areas, except on very steep slopes.

Bench Terracing: On steeply sloping lands, the slopes where such terraces are found useful vary from 6 to 30 per cent. Bench terraces with 100 m length, longitudinal grades in the range of 0.2 to 0.8 per cent are recommended for Alfisols of high rainfall regions.

Ploughing: Ploughing across the slope and growing low value crops in catchment areas, the ploughing of deep soils should be done once in three to four years immediately after *rabi* crops. The light, shallow and medium soils should be hoed instead of ploughing which help to receive and retain moisture.

Reclaiming Problem Soils: Reclaimation of Acidic Alkaline and Saline soils should be reclaim by adding lime, gypsum, sulphur, or pyrites respectively. Growing high value crops in level run-off concentrated strips and incorporating a liberal quantity of organic matter.

Maintenance of Soil Fertility and Yield Stability

Dryland areas have low yields and high yield fluctuations. The maintenance of soil fertility is a problem in such areas as for a considerable period of the year the soil remains uncropped and there is a loss of plant nutrients, loss of the fertile surface due to erosion leads to a decline in soil fertility to build up soil fertility and reduce the fluctuation of crop yield.

a. the combined use of farmyard manure and green manure with inorganic fertilizer,
b. the incorporation of crop residues,
c. the inclusion of fodder-legumes/legumes in the cropping system/ rotation,
d. the use of bio-fertilizers,
e. suitable methods of application of fertilizers, and mulching ,
f. checking loss of surface soil by using soil stabiliser and chemicals e.g. Na_2,Co_2,polythene sheet, mulches and plastic sheets.
g. developing small agricultural watersheds for run off collection and recycling for life saving irrigation to crops in moisture stress,
h. allowing a portion of the holding as chemical or legume fallow which on cultivation provides a substantial yield during famine,
i. adopting alternate land use planning in conjunction with regular cropping to improve the income of the farming families from fibre, fuel, fruit, furniture-timber, fodder and farm animals along with food.

Conservation of soil and water in dryland areas are inter-related and where one is tried the other is also achieved.

The World's Drylands

Definition of Drylands

Depending on definitions, about 47 percent of the surface of the earth can be classified as dryland (UNEP, 1992). Although there is no clear boundary, drylands are considered to be areas where average rainfall is less than the potential moisture losses through evaporation and transpiration. According to the World Atlas of Desertification (UNEP, 1992), drylands have a ratio of average annual precipitation (P) to potential evapotranspiration (PET) of less than 0.65.

Where the water deficit prevails throughout the year, drylands are classified as extremely arid or hyperarid, whereas when it occurs for most of the year they are arid and semi-arid regions. Aridity is assessed on the basis of climate variables (so-called aridity index), or according to FAO on the basis of how many days the water balance allows plant growth (growing season). The aridity index uses the P/PET to classify drylands into hyperarid, arid, semi-arid and dry subhumid.

The negative balance between precipitation and evapotranspiration results in a short growing season for crops (usually less than 120 d). For CS purposes, drylands are also considered to include arid, semi-arid and dry subhumid areas. Hyperarid regions are not considered as there is no crop growth unless under irrigation.

Droughts are characteristic of drylands and can be defined as periods (1 - 2 years) where the rainfall is below the average. Droughts that persist for a decade or more are called desiccation, which can have disastrous consequences for land productivity and vegetation loss. Drought preparedness and risk mitigation are essential for the proper management of dryland areas. Populations living in these regions have been developing strategies to cope with them. These measures include: strengthening indigenous strategies to cope with drought; supporting the development and adoption of resource management practices that will protect and improve productivity, thereby increasing the resilience of agricultural systems; reducing fluctuations in prices of livestock and grains during drought periods through expanding market size and reducing transaction costs; developing a set of warning indicators; and setting aside drought grazing reserves or strategic water reserves (Øygard, Vedeld and Aune, 1999).

Land Degradation in Drylands

Desertification results from the degradation of the natural ecosystems in drylands and constitutes a major global problem (UNEP, 1992). It is defined by the CCD as "*Land use degradation in arid, semi-arid and dry humid areas resulting from various factors, including climatic variation and human activities*". The degradation can be:

Table: *Dryland categories according to FAO (1993) classification and extension (UNEP, 1992)*

Classification	*P/PET (UNEP, 1992)*	*Rainfall (mm)*	*Area (%)*	*Area (Bha)*
Hyperarid	< 0.05	< 200	7.50	1.00
Arid	0.05 < P/PET < 0.20	< 200 (winter) or <400 (summer)	12.1	1.62
Semi-arid	0.20 < P/PET < 0.50	200 - 500 (winter) or 400 - 600 (summer)	17.7	2.37
Dry subhumid	0.50 < P/PET < 0.65	500 - 700 (winter) or 600 - 800 (summer)	9.90	1.32
TOTAL			47.2	6.31

Bha = 10^9 ha.

- physical — mainly driven by climate factors such as floods and droughts that cause soil erosion (by wind and water),
- chemical — generally in the form of salinization (in irrigated lands),
- biological — mainly as a result of the oxidation of topsoil organic matter in dryland.

The main consequences of land degradation are: the chemical degradation of the soil; loss of vegetation cover; loss of topsoil infiltration capacity; reduction in soil water storage; loss of SOM, fertility and structure; loss of soil resilience; loss of natural regeneration; and lowering of the water table. Soil degradation affects about onefifth of arid zones, mostly on semi-arid margins where cultivation take place. Land degradation may have a significant impact on climate. The loss of plant cover can alter the surface energy balance. Atmospheric dusts from deserts modifies the scattering and absorption of solar radiation (Kassas, 1999). Although uncertainty exists with regard to the causes of climate change and global warming and the possible consequences, there is agreement that some impacts are probable. For example, temperature increases will affect evapotranspiration, which will be most significant in places where the climate is hot. Predictions about the quantity and distribution patterns of rainfall in these regions are uncertain, but the Intergovernmental Panel on Climate Change indicated that semi-arid regions are among those most likely to experience increased climate stress (IPCC; 1990). Furthermore, climate change may have unpredictable and perhaps extreme consequences with respect to the frequency and intensity of precipitation and temperature variability for semi-arid regions.

The extension of degraded lands according to cause. One of the problems of assessing the extent of desertification and the measures to prevent it, is the lack of reliable and easily measured land quality indicators.

The Land Degradation Assessment in Drylands project, initiated by FAO, focuses on the development of a detailed methodology for the assessment of land degradation in an area that covers as much as half of the global land surface (FAO, 2002a, 2003).

Several estimates exist for the extent of desertification. According to the Global Assessment of Human and Induced Soil Degradation methodology, the land area affected by desertification is 1 140 000 000 ha, which are similar to the UNEP estimates.

According to UNEP (1991a), when rangelands with vegetation degraded are included (2 576 000 000 ha), the percentage of degraded lands of the drylands is 69.5 percent (5 172 000 000 ha). According to Oldeman and Van Lynden (1998), the degraded areas for light, moderate and severe degradation are 489 000 000, 509 000 000 and 139 000 000 ha respectively.

Table: *Degraded lands per continent*

Cause	*Africa*	*Asia*	*Oceania*	*Europe*	*North America*	*South America*
	(million ha)					
Deforestation	18.60	115.5	4.20	38.90	4.30	32.20
Overgrazing	184.6	118.8	78.50	41.30	27.70	26.20
Agricultural	62.20	96.70	4.80	18.30	41.40	11.60
Over exploitation	54.00	42.30	2.00	2.00	6.10	9.10
Bio-industrial	0.00	1.00	0.00	0.90	0.00	0.00
Total degraded	319.4	370.3	87.50	99.40	79.50	79.10
Total	1286	1671.8	663.3	299.6	732.4	513.0

Source: UNEP (1997).

Estimates of rates of current desertification vary considerably mainly because of the lack of quantitative criteria for defining degradation. UNEP (1991a) distinguished between land degradation and vegetation degradation. The degradation of vegetation in rangelands can take place with or without soil degradation. UNEP (1991) estimates the annual rate of desertification to be 5 800 000 ha or 0.13 percent of the dryland in mid-latitudes. However, although desertification is a problem in drylands, drylands have a high degree of resilience to human interventions. Dryland populations have developed welladapted and efficient resource management practices. Therefore, the participation of dryland communities is crucial to improving dryland management.

If the policies and practices of donors are to succeed, they must be based on the knowledge, experiences, aspirations, priorities and decisions of the people living in drylands.

Table: *GLASOD estimates of desertification (excluding hyper dry areas)*

Land type	*1. Area (Bha)*	*Type of soil degradation*	*2. Area (Bha)*
Degraded irrigated lands	0.043	Water erosion	0.478
Degraded rainfed croplands	0.216	Wind erosion	0.513
Degraded range- lands	0.757	Chemical degradation	0.111
(soils and vegetation)		Physical degradation	0.035
Total land area	1.016	Total land area	1.137

Bha = 10^9 ha.

Sources: 1. UNEP (1991b). 2. Oldeman and Van Lynden (1998).

Table: *Rates of land degradation in mid-latitudes drylands*

Land use	*Total land area (Mha)*	*Rate of desertification*	
		Mha/y	*Percent of total/y*
Irrigated land	131	0.125	0.095
Rangeland	3 700	3.200	0.086
Rainfed cropland	570	2.500	0.439
Total	4 401	5.825	0.132

Mha = 10^6 ha.

Source: (UNEP, 1991a).

Desertification can be prevented through a proper management of the land to ensure sustainable development of its resources. In 1994, the United Nations agreed on the CCD by developing specific country action plans.

Strategies for desertification control include: establishment and protection of vegetation cover to protect soils from erosion, controlled grazing; improved water conservation by residue management and mulching to help decrease water losses by runoff and evaporation; supplemental irrigation; soil fertility management which enhances biomass productivity; increased water use efficiency; and improved soil quality; improved farming systems that include crop rotations; fallowing; agroforestry; and grazing management (Lal, 2001b). All these strategies increase CS in soils.

Depending on land-use, desertification is manifested in different ways:

Irrigated farmlands:	Excessive irrigation and inefficient drainage leads to waterlogging and salinization;
Rainfed farmlands:	Soil erosion, loss of organic matter and nutrients;
Rangelands:	Reduction in plant productivity, invasion of unpalatable species.

Desertification affects more than 100 developed and developing countries in all continents (UNEP, 1997). Some 200 million people are believed to be affected directly by desertification and more than 1 000 000 000 people at risk. The future sustainability of dryland ecosystems and the livelihoods of people living in them depend directly on the actions taken for land-use management. These activities should include soil and water conservation for improved land-use management practices and farming systems, taking into account health, social and economic issues when developing strategies and policies to improve land management.

Distribution of Drylands

Most arid land areas of the world occur between the latitudes of 20° and 35°. The main semi-arid areas occur on each side of the arid zone and include Mediterranean-type and monsoonal-type climates. Mediterranean climates are characterized by cold wet winter and dry hot summers whereas monsoonal-type climates have hot wet summers and warm dry winters. Another type of dryland is the cold desert, which generally occurs in high-altitude continental areas

Drylands occupy 47.2 percent of the world's land area, or 6 310 000 000 ha across four continents: Africa (2 000 000 000 ha), Asia (2 000 000 000 ha), Oceania (680 000 000 ha), North America (760 000 000 ha), South America (56 000 000 ha) and Europe (300 000 000 ha) (UNEP, 1992) in more than 110 countries. About 2 000 000 000 people live in drylands (UNEP, 1997), in many cases in poor conditions. The hyperarid zones extend mostly across the Saharan, Arabian and Gobi deserts and have only localized population around valleys such as the Nile Valley and the Nile Delta. The arid zones cover about 15 percent of the land surface. The annual rainfall in these areas is up to 200 mm in winter-rainfall areas and 300 mm in summerrainfall areas. Interannual variability is 50 - 100 percent. Africa and Asia have the largest extension of arid zones, they account for almost four-firths of hyperarid and arid zones in the world.

Semi-arid zones are more extensive and occur in all the continents, and cover up to 18 percent of the land surface. They have highly seasonal rainfall regimes and a mean rainfall of up to 500 mm in winter-rainfall areas and up to 800 mm in summer-rainfall areas. With an interannual variability of 25 - 50 percent, grazing and cultivation are both vulnerable, and population distribution depends heavily upon water availability.

Soil and Vegetation of Drylands

As discussed above, the soils of drylands are characterized by frequent water stress, low organic matter content and low nutrient content, particularly nitrogen (N) (Skujins, 1991). Although dryland vary considerably, they are mostly Aridisols (2 120 000 000 ha) and Entisols (2 330 000 000 ha). Other soils include: Alfisols (380 000 000 ha), Mollisols (800 000 000 ha), Vertisols (210 000 000 ha) and others (470 000 000 ha) (Dregne, 1976). Whatever their type, soils are the basic resource of drylands as they provide the medium in which plants grow, and their properties, such as texture and waterholding capacity, determine the proportion of rainfall available for plant growth. Low organic matter content, low germination and high seedling mortality are the main causes of very low plant productivity.

***Table:** The global dryland areas by continent*

Continent	*Extension*			*Percentage*		
	Arid	*Semi-arid (million ha)*	*Dry subhumid*	*Arid*	*Semi-arid*	*Dry subhumid*
Africa	467.60	611.35	219.16	16.21	21.20	7.60
Asia	704.30	727.97	225.51	25.48	26.34	8.16
Oceania	459.50	211.02	38.24	59.72	27.42	4.97
Europe	0.30	94.26	123.47	0.01	1.74	2.27
North/central America	4.27	130.71	382.09	6.09	17.82	4.27
South America	5.97	122.43	250.21	7.11	14.54	5.97
Total	1 641.95	1 897.74	1 238.68			

The vegetation supported by these soils ranges from barren or sparsely vegetated desert to grasslands, shrublands and savannahs, croplands and dry woodlands. Forest vegetation is usually poor, and is at low density with species adapted to arid soils and with a high water-use efficiency. Perennial vegetation varies considerably and tends to be sparse and patchy. Plants that have adapted to drylands survive irregular rainfall, high solar radiation and drought periods. Plants protect the soil surface from wind and water erosion. Removal or loss of vegetation cover results in an increased risk of soil erosion and degradation.

***Table:** Typical crops under rainfed conditions*

Classification	*Length of the growing season*	*Typical crops*
Hyper-arid	0	No crop, no pasture
Arid	1 - 59	No crops, marginal pasture
Semi-arid	60 - 119	Bulrush millet, sorghum, sesame
Dry subhumid	120 - 179	Maize, bean, groundnut, peas, barley, wheat, teff (suitable for rainfed agriculture)

FAO, 1993.

***Table:** Percentage land uses in arid regions in 1980*

Nomadic pastoralism	41
Ranching	25
Rainfed agriculture	12
Hunting, fishing, gathering	3
Irrigated agriculture	2
mostly unused	16

Source: Heathcote (1983).

The predominant land uses of the drylands are pastoralism and subsistence food production. Cereals produced in drylands include wheat, barley, sorghum and millet and pulses such as chickpea, lentils, peas and groundnuts. Less important are oil crops (rape and lindseed) and a wide range of fruits, vegetables, herbs and spices. Pastoralism is widespread and highly mobile. Food production is mainly from smallholding rainfed systems for subsistence or local consumption and markets. Natural woodlands are used for fuel wood, and efforts are ongoing to extend the forested areas for fuel and for CS.

The major constraint on agricultural development is low and highly variable rainfall and the consequent high risk for agriculture and animal husbandry. Traditional systems of rainfed cropping have evolved for thousands of years. Several general strategies have been developed to cope with low and erratic rainfall. Rainfed agriculture is generally practised in areas with a reasonable amount of rain and where soils are relatively deep. Drier regions are generally used for livestock grazing, with regular seasonal movements. Normally, several crops are sown to reduce the risk of total crop failure. Varieties that are resistant or adapted to drought are used. Long fallows are used to prevent stress on the land. During the fallow periods, soils are protected by a vegetation cover that provides nutrient and organic matter to the soils. Many pastoralists and sedentary farmers work together by exchanging crops and meat.

Characteristics of Drylands that Affect Carbon Sequestration

Dryland environments are characterized by a set of features that affect their capacity to sequester C. The main characteristic of drylands is lack of water. This constrains plant productivity severely and therefore affects the accumulation of C in soils. The problem is aggravated because rainfall is not only low but also generally erratic. Therefore, good management of the little available water is essential. In addition, the SOC pool tends to decrease exponentially with temperature (Lal, 2002a).

Consequently, soils of drylands contain small amounts of C (between 1 percent and less than 0.5 percent) (Lal, 2002b). The SOC pool of soils generally increases with the addition of biomass to soils when the pool has been depleted as a consequence of land uses (Rasmussen and Collins, 1991; Paustian, Collins and Paul, 1997; Powlson, Smith and Coleman, 1998, Lal, 2001a). Soils in drylands are prone to degradation and desertification, which lead to dramatic reductions in the SOC pool.

A good overview of the extent of land degradation in different dryland regions of the world is given in Dregne (2002). However, there are also some aspects of dryland soils that work in favour of CS in arid regions.

Dry soils are less likely to lose C than wet soils (Glenn *et al.*, 1992) as lack of water limits soil mineralization and therefore the flux of C to the atmosphere.

Consequently, the residence time of C in drylands soils is long, sometimes even longer than in forest soils. The issue of permanence of C sequestered is an important one in the formulation of CS projects. Although the rate at which C can be sequestered in these regions is low, it may be cost-effective, particularly taking into account all the side-benefits resulting for soil improvement and restoration.

Soil-quality improvement as a consequence of increased soil C will have an important social and economic impact on the livelihood of people living in these areas.

Moreover, given the large extent of drylands, there is a great potential for CS. The potential offered by drylands to sequester C is large, not only because of the large extent, but because historically, soils in drylands have lost significant amounts of C and are far from saturation. Because of all of these characteristics, any strategy to re-establish SOM in these regions is particularly interesting (Box 1).

Dryland Characteristics That Affect Cs

Unfavourable

- Lack of water
- Low and erratic rainfall
- Generally high temperatures
- Low productivity
- Low SOM (0.5-1percent) and nutrient content
- Prone to soil degradation and desertification

Favourable

- Residence time of SOM is long
- They occupy more than 43 percent of the earth's surface
- As a consequence of historic carbon loss they are far from saturation
- Soil quality improvement through CS will have large economic and social impac

Desertification and Carbon Sequestration

The effects of desertification on soil quality include:

- loss in soil aggregation
- decrease in water infiltration capacity
- reduction in soil water storage · increase in erosion potential
- depletion in SOM, dif.culty in seed germination
- disruption of biogeochemical cycles C, N, phosphorous, sulphur alterations in water and energy balance
- loss of soil resilience

All of these effects accentuate the emission of CO_2 to the atmosphere. Lal (2001c) estimated the C loss as a result of desertification. Assuming a C loss of 8 - 12 Mg C/ha (Swift *et al.*, 1994) on a land area of 1 020 000 000 ha (UNEP, 1991a), the total historic C loss would amount to 8 - 12 Pg C. Similarly, vegetation degradation has led to a C loss of 4 - 6 Mg C/ha on 2 600 000 000 ha, adding up to 10 - 16 Pg C. The total C loss as a consequence of desertification may be 18 - 28 Pg C. Assuming that two-thirds of the C lost (18 - 28 Pg) can be resequestered (IPCC, 1996) through soil and vegetation restoration, the potential of C sequestration through desertification control is 12 - 18 Pg C (Lal, 2001c). These estimates provide an idea about the loss of C as a result of desertification and the potential for CS through the restoration of soils in drylands.

Opportunities for improved land management as well as increasing CS should be developed in these areas. Agricultural systems contribute to carbon emissions through the use of fossil fuels in farm operations and through practices that result in loss of organic matter in soils. On the other hand, farming systems can offset carbon losses when accumulating organic matter in the soil, or when aboveground woody biomass is increased, which then acts either as a permanent sink or used as an energy source that substitutes fossil fuel. The potential for global benefits, as well as local benefits, to be obtained from increased CS in drylands should be an additional incentive for stronger support for reforestation and agriculture in drylands.

Although drylands have been studied (Heathcote, 1983; Thomas, 1997a, 1997b), the impact of desertification on the global carbon cycle and the potential impact of desertification control on CS in dryland ecosystems have not been widely investigated. There are few case studies, and little information. Consequently, there is little scientific evidence on the impact of desertification on carbon emission to the atmosphere. The aim here is to assess the state of knowledge, and the potential of different measures to increase CS.

Farming Systems in Drylands

According to FAO (2001a), a farming system is defined as "*a population of individual farm systems that have broadly similar resource bases, enterprise patterns, household livelihoods and constraints, and for which similar development strategies and interventions would be appropriate*". Depending on the scale of the analysis, a farming system can encompass a few dozen or many millions of households. The understanding of the major farming systems in drylands provides the necessary framework for the development of agricultural strategies and interventions. Based on the classification of farming systems of developing regions specified by FAO (2001a), most of the farming systems in drylands fall into the category of rainfed farming systems in dry lowpotential areas. These systems are characterized by mixed crop - livestock and pastoral systems merging into sparse and often dispersed systems with very low current productivity or potential because of extreme aridity or cold.

Understanding the world of smallholders in dryland environments is the key to designing appropriate and successful CS activities. It is important to understand that CS for poverty alleviation must be much broader in terms of the range of both practices and benefits (i.e. not only in monetary terms) than similar schemes in commercial agriculture and forestry.

There are various potential partners, or target groups, for CS programmes in drylands. From a purely scale-driven perspective, large-scale capital-intensive agriculture might be the most attractive. However, from a biophysical standpoint, systems that use significant quantities of fertilizers or that depend heavily on fossil fuel to supply irrigation water should not generally be considered because they are usually net carbon emitters. Only if a switch from high fertilizer and fossil-fuel dependence to more carbon-friendly inputs, technologies or land use is foreseeable in the short-run, should current large-scale agriculture be considered. There are some systems of capital intensive land use, such as the mechanized farming schemes in eastern Sudan (where large areas of land have been severely degraded) that offer great potential for soil CS if rehabilitated through low-intensity land use.

Apart from these technical reasons, large-scale, capital-intensive agriculture systems are probably not potential partners for soil CS because the small additional income that sequestration might bring would be unattractive in comparison with profits from other sources, many of which depend on carbon-emissive techniques.

Thus, the main target groups of soil CS in degraded agro-ecosystems are primarily small-scale, resource-poor farmers in uncertain and risk-prone environments for whom anticipated benefits could constitute an enhancement of their livelihood. Reference to these groups of farmers is made as smallholders. They depend on lowinput, subsistence-based agriculture, and they are usually characterized by diversity, variability and flexibility (Mortimore and Adams, 1999).

Characteristics of Smallholder Agriculture

The primary characteristics of smallholder agriculture in semi-arid developing countries are its diversity in space, its variability through time, and its multidimensionality in terms of the ways it operates and survives (Mortimore and Adams, 1999). This is largely because dryland smallholders must be highly responsive to a varied, changeable and hazardous environment. Thus, their operations are very different from those of large-scale farms driven by commercial goals, equipped with credits and efficiencyoriented technologies and covered by insurance systems against hazards and losses. This diversity, variability and multidimensionality means that each particular system must be approached with careful attention to its unique mix of characteristics.

Another important characteristic of smallholders, which also differentiates them from commercial farmers, is that few are motivated solely by the goal of agricultural profit. Instead, smallholders pursue basic subsistence and survival goals, balancing daily risks and opportunities directly through their livelihood options and management practices rather than through external institutions (Collinson, 2000). Many smallholders have deep attachment to their land, which they continue to farm, even when profitability is low, for reasons such as maintaining tenure, and maintaining family ties. At the same time, many have additional, and often higher, incomes from non-agricultural sources. These include: petty trading; the gathering of wild produce, including firewood; labouring; and remittances from family members. The result is "multi-enterprise production units" (Hunt, 1991).

Smallholders are further differentiated from high-input commercial farmers by their need to manage multiple risks. Almost all of their inputs and outputs are subject to large variation and uncertainty, such as labour, which is often the most critical variable. Another critical risk arises from the high variability in rainfall, which itself has two major consequences as far as sequestration is concerned. One is variation in the timing of bioproductivity, which means that planting and harvesting (and most other agricultural and non-agricultural activities) may have to be readjusted rapidly, sometimes within a season, and often between seasons. For example, fallows that appeared secure for years may have to be cleared after a particularly poor season. The other consequence is variability between fields, some of which may receive sufficient rainfall, and some of which may not. There are other risks that have similar consequences. These include: attacks by pests (against which pesticides are too expensive); illness, resulting in the unavailability of labour at some critical point in the season; and variability regarding prices of inputs such as seed, labour, food, and of outputs, mainly crops.

According to Mortimore and Adams (1999), smallholder responses to these various constraints follow three key avenues: (i) diversification of natural, economic, technical and social resource endowments with the underlying rational to spread risks as efficiently as possible; (ii) flexibility in day-to-day management of these resources in the form of active decisions to cope with and adapt to short-term variability; and (iii) adaptability over the longer term, perceived as cumulative and purposeful decisionmaking that will result in new or altered systems or livelihood pathways. When spreading risks, it is important for farmers to have a mix of products where both the type of products and the price of these products are independent of each other, a criteria that potentially applies very well to CS.

A further characteristic of smallholder agriculture is variable access to resources of all kinds. Within a village, some have ready access, and others have less access, to: secure landholdings; wild produce, such as fuelwood; credit; hired labour; livestock; and markets. Access also varies between villages and between countries. The implications of such uneven access to resources for sequestration schemes are discussed below.

Finally, these agricultural systems are and have long been undergoing continual change in response to environmental and social changes. Dry environments are now widely recognized as having a complex history of change, based on nonequilibrium dynamics rather than predictable, gradual and linear change (Leach and Mearns, 1999; Scoones, 1999; Scoones, 2001), sometimes referred to as event-driven systems (Reenberg, 2001; Sorbo, 2003). Thus, agricultural systems have had to adapt continuously to environmental conditions and to changing political and economic processes. In the lifetime of a soil CS scheme, one could expect many changes in the configuration of the agricultural landscape, apart from the changes that the project itself might bring. Planning in such an environment will be challenging. Instead of simplified and standardized approaches and predefined technical solutions, CS schemes in these systems will need to offer a range of technological and management options from which farmers can choose according to their needs.

Examples of Smallholder Farming Systems

Within this broad description of the characteristics of dryland smallholder agriculture, there are various farming systems. These are systems such as annual croplands, plantations, forests, savannahs, natural pastures, fallow lands and vegetable gardens. Within each, there is a specific interaction between crops, livestock and trees, and between cultivated and non-cultivated land (FAO, 2000a).

Farming systems in drylands range from shifting cultivation embedded in extensive wooded grasslands to intensive smallholder farming where all land is under cultivation and the integration between cropping and animal husbandry maximized. However, these two extremes should not be understood as fixed points along an axis of agricultural development, but rather as examples of "pathways" of agricultural and environmental change (Scoones, 2001) that are possible both between and within sites. Such pathways of change reflect farmers' livelihoods, constraints and opportunities within a historical context.

Agricultural Intensification

Intensification, as defined by Tiffen and Mortimore (1993), implies "increased average inputs of labour or capital on a smallholding, either on cultivated land alone, or on cultivated and grazing land, for the purpose of increasing the value of output per hectare". Intensification takes many forms, which can be classified in many ways. In the case of smallholder dryland farming systems, intensification tends to be related to increased local labour inputs per hectare and low-cost technologies rather than capitalintensive innovations. Mortimore and Adams (1999) describe such intensification as an "indigenous and adaptive process" whose path can be reconstructed through historical analyses.

There are many examples of such "indigenous" intensification. In rainfed farming systems, intensification often occurs as a consequence of growing population pressure. In many places, fallow periods have become shorter and shorter and eventually even abandoned. All fields may then be under cultivation and soil fertility is maintained by greater labour intensity. Techniques may include: intercropping with N-fixing legumes; time-intensive weeding and harvesting; the utilization of manure and mulch; and the protection of certain tree species. Crop rotation is practised where possible to ensure differential nutrient use and uptake between crops, such as millet and sorghum, and Nfixing crops, such as groundnuts and cowpeas. Trees, especially those known for their N-fixing and soil-restoring capacities, are protected. The application of manure, either from cattle or small ruminants, is a key element. In order to maintain supply in the face of increasing land scarcity, herds must be managed more intensively, e.g. feeding them with agricultural residues and weeds.

In dryland areas where sufficient surface water is available, irrigation has been a key method of intensifying land-use systems since ancient times. It requires supplies of water and of energy to take the water to the fields and gardens. The water may come from streams, rivers, springs and wells. Streams and rivers may be range from small ephemeral water courses, as in many parts of central Asia, to major rivers such the Nile, Niger, Amu-Darya, Hwang He and Indus.

Where there is a good "head" of water, as in mountainous terrain (e.g. in parts of highland Yemen and Oman), or in the large systems on floodplains, as of the Indus and Nile rivers, the water may be taken by gravity in small channels to the fields or gardens. Where the river flows in a gently sloping floodplain, the methods raising it to the fields of small schemes are much the same as from wells: animal or human-driven devices such as saqqias, Archimedes screws and shadufs. The qanat systems, which are particularly well developed in Iran and neighbouring areas, but are also found in other parts of Asia and in north Africa, are more elaborate, involving wells from which water is channelled underground, and fed to the fields by gravity. Another ancient system, which has seen major expansion and development in recent years, is water harvesting (or runoff agriculture). In this system, highly intermittent runoff is concentrated and then held in shallow troughs, where it is usually used for tree-crops.

Extensive Land Use

In areas where both population densities and rainfall are low, patterns of extensive land use predominate either as a longer-term system state or a more recent major pathway of change (Mortimore and Adams, 1999). The latter is true for some areas in central Senegal where entire compounds have migrated recently to the city of Touba, leaving relatives and neighbours with more land available than in previous decades (Tschakert and Tappan, 2004). As land scarcity does not represent a constraint in this case, fallow lands constitute an important element of the farming system, allowing for short- and medium-term soil regeneration. In general, field sizes are significantly larger than in areas under intensification. Given the amount of land available to individual households, manure is generally only used for fields that are under continuous cultivation, primarily those adjacent to the settlements and others in close proximity. Remote fields and those left fallow are accessible to grazing animals all year round. Unlike animals in intensified systems, herds are not forced to leave for transhumance and thus contribute to a continuous flux of organic matter input. Weeding and harvest activities might occur with less intensity, while more agricultural residues are left on the fields.

Agroforestry may play an important part in these extensive systems. One example is the Sudanese system of gum arabic production, where a tree that regrows on fallow land is a major source of income for smallholders (Elmqvist and Olsson, 2003). In other long fallows, trees that yield other useful products, such as fruits, nuts, fibre and medicines, are planted. Trees also provide an important source of emergency food.

The above examples of intensive and extensive farming systems illustrate that a context-specific approach based on multiple pathways of change offers useful guidelines for potential CS schemes. Project design and implementation should start with a local understanding of environmental change and its underlying processes. The next step is to identify positive pathways of change at the local level and then finally to assess opportunities to encourage such pathways on a larger scale.

Soil Fertility Management

The concept of CS in degraded agro-ecosystems is typically based on two assumptions. The first is that any improvement in soil fertility management and land use will result automatically in higher amounts of C sequestered from the atmosphere and stored in soils. The second is that local smallholders and herders, who are anticipated to be the prime beneficiaries of planned interventions, need to be made aware of and trained in such improved management practices.

Given the complex, diverse and dynamic world of smallholder farming in dryland environments, these two assumptions seem oversimplified. In general, proposed management practices and land-use options merely reflect the most efficient technical options, focusing on achieving an optimal agronomic situation. However, as illustrated above, smallholders are more concerned about day-to-day risk management and longer-term adaptive strategies than the achievement of an assumed new equilibrium. Opportunistic farming is all about spreading risk, an adaptive process during which both losses and gains occur, often intentionally. Pure "efficiency would leave no room for flexible maneuver" (Mortimore and Adams, 1999).

What in fact constitutes "improved" soil fertility management or land-use options might be understandable only from a holistic farming-system research approach. Farmers who have developed highly dynamic and flexible soil-fertility management practices to cope with variability and uncertainty are often in the best position to bring this holism to a development project. Although farmers often have much experience in deliberating technologies within a much broader framework of "real life", they are most often considered as passive recipients of outside assistance rather than key resources in the process itself.

Thus, a first step towards linking soils and C to people is to investigate practices that smallholders in dryland environments currently know and use, to understand their underlying rationale as well as driving factors for change, and to identify examples of positive pathways of change that could be replicated on a larger scale (Tschakert and Tappan, 2004).

Soil fertility management practices can be grouped according to the movement of nutrients into, within, and out of a system. Here, practices are sorted into four groups (Hilhorst and Muchena, 2000): (i) adding nutrients to the soil; (ii) reducing losses of nutrients from the soil; (iii) recycling nutrients; and (iv) maximizing the efficiency of nutrient uptake. The examples below are based mainly on the Senegal and Sudan case studies.

Adding Nutrients to the Soil

Fallowing: Fallowing is a well-known practice for replenishing nutrients in soils. Ideally, fallow periods are rotated with cropping periods, allowing the land to recover from years of cultivation. However, in many parts of the world's drylands, both fallow area and duration have decreased over time. Most often, this decline is caused by increasing population pressure, the introduction of modern agricultural machinery, such as the plough, and periods of droughts, or a combination of all three. Some believe that this process is reaching crisis proportions. Today, in many drylands, fallow duration is reduced to only one year. In areas with severe land scarcity, it has disappeared altogether. As a consequence, farmers have shifted to other soil-fertility management practices, such as manuring and composting or they continue to farm with exceptionally low and decreasing yields. At the same time, less land in fallow also means reduced grazing possibilities or less fodder for animals, thus reducing the amount of manure that can be produced (Breman, Groot and van Keulen, 2000). Nevertheless, in areas with less population pressure, fallowing still constitutes an important option for soil-fertility management. This is particularly true for countries where structural adjustment packages have been implemented and subsidies for fertilizers removed.

Stubble Grazing

Many farming systems include the grazing of animals on fields immediately after crop harvests. Animals graze stubble and stalks left on the field, while soils benefit from faeces deposition throughout the duration of the practice. Depending on the size and the forage situation of a field, as well as the overall number of animals, livestock is usually kept for 1 - 7 months on the same field, where it is rotated between different parts during shorter time intervals.

Overall, the gain in organic matter from stubble grazing can be substantial. In the Sahel, deposition of droppings ranges from 1 tonne/

ha to 50 tonnes/ha depending on the time that animals are kept on the same field (Sagna-Cabral, 1989; Garin and Faye, 1990; Hoffmann and Gerling, 2001). However, direct exposure to the elements can reduce the nutrient value of dung and droppings considerably. Although stubble grazing has a long tradition in drylands, increasing land scarcity, limited purchasing power among many smallholders, and increased risks of animal theft in many areas have contributed to a general decline in herd sizes and, in some cases, led to the abandonment of stubble grazing altogether.

Inorganic Fertilizers

The use of inorganic fertilizers has been one of the most widely promoted means for increasing production since the early twentieth century. In many of the drylands of the developing world, this kind of fertilizer was subsidized and made available to farmers with the aid of government and the support of non-governmental organizations (NGOs). Under structural adjustment programmes, subsidies were often removed and, hence, fertilizers became increasingly expensive for farmers. As is shown in the Senegal example, the use of fertilizers decreased in the 1990s. From a CS point of view, the use of synthetic fertilizers does not result in any net gain of carbon fixation (Schlesinger, 1999). The emission of CO_2 during the manufacture, transport and application of the fertilizers offsets any gain in biological production.

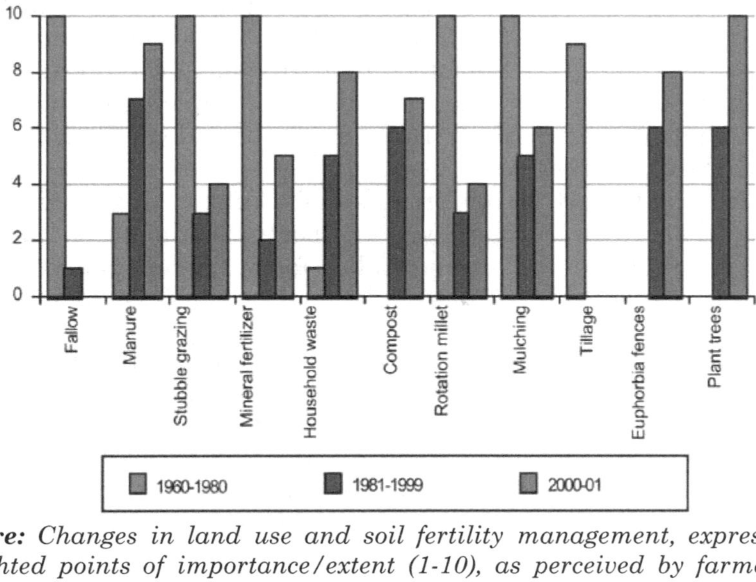

Figure: *Changes in land use and soil fertility management, expressed in weighted points of importance/extent (1-10), as perceived by farmers in an intensified farming system in Senegal*

Source: Tscharkert, field work, 2001.

Crop Rotation and Association

The practice of crop rotation and association, especially where it involves both cereals and legumes, is well known among farmers as a soil-fertility management practice. In many places, N-fixing crops include beans and groundnuts. However, in farming systems where land scarcity has become a limiting factor, priority is often given to cereals. Moreover, the availability of seeds for legumes might be dependent on state subsidies or credits, as was the case with groundnuts in the Sahel.

Reducing Losses of Nutrients From the Soil

Woody Vegetation: Trees can be an important component in many agro-ecosystems. With their deep and extensive root systems, they can capture nutrients not accessible by crops and make them available to crop production again through litter fall. From a CS point of view, not only do the trees store C in their aboveground biomass, they also contribute to belowground biomass through their root systems and their input of litter to the soil (branches and leaves). Of particular use are leguminous (N-fixing) trees, among which *Faidherbia albida* and *Acacia senegal* are two of the most appreciated. Trees may also play a role in reducing nutrient losses from wind erosion. Special consideration must be given to the use of biofuel instead of fossil fuel.

Living hedges and fences can capture silt and clay particles suspended in the air and could, hence, locally increase the clay content of the soil, a factor beneficial for CS (El Tahir and Madibo, in press). Litter produced by woody plants is beneficial because of its higher content of polyphenols (lignines and tannins), which decreases the decomposition rate (Abril and Bucher, 2001), when compared with grasses and annual herbs.

Erosion Control

Erosion and subsequent transport and deposition have a complex relationship to soil carbon storage. Where water erosion dominates, a high proportion of soil C may be washed into alluvial deposits close to the erosion site, and stored there in forms which decay more slowly than in the parent soils. Therefore, this kind of erosion may have a positive effect on soil CS. Erosion does not always decrease productivity, but if it could be shown to do so, it would be perverse to favour decreased productivity for a mediumterm and perhaps one-off gain in sequestered C. The same arguments probably do not apply where wind erosion is the main erosional process, for organic matter is usually blown great distances and dispersed to places where it may decay rapidly and release its C. Management options that increase the amount of live and dead biomass left in agricultural areas decrease erosion in general while simultaneously increasing the C input to the soil (Tiessen and Cuevas, 1994).

Field Clearing and Weeding

Clearing the fields from weeds before planting, as well as weeding during the cropping season, is an important practice for reducing competition between the crop and weeds. However, from the point of view of soil fertility and CS, it is important to recycle as much of the weeds back into the soil as possible. Selective clearing and weeding implies that only weeds competing directly with the crop are removed while others remain in the field.

Recycling Nutrients

Manure: Spreading manure from livestock that is kept within or close to the compounds is one of the most widespread soil fertility management practices. Farmers are well aware of the fertilizing effects of manure but they also appreciate it for the fact that it stabilizes sandy topsoil and reduces wind erosion. Farmyard manure (FYM) and manure produced in pens is usually of higher quality than dung and droppings left on fields by grazing animals. It may be combined with agricultural residues, household waste and ashes accumulating within a household. The most limiting factor of the usage of manure, in addition to lack of animals, is the lack of transporting material, often resulting in well-manured fields closer to the homestead rather than on more remote fields.

Crop Residue Management

Crop residues such as stalks and hay can be left on or returned to the field in the form of mulch or ploughed under at the end of the cropping season. However, in most dryland farming systems, crop residues are contested and are removed after harvest either as fodder, construction material, fuel or litter for composting. What remains on the field is often burned before the next cropping season. In some cases, crop residues are also sold at the local market, generating additional income.

Management of Other Organic Matter

Household waste, fish scales, ashes, leaf litter, prunings, and surplus crop residues are also used to increase soil fertility. Often, these additional inputs are accumulated within the homestead, sometimes added on to manure piles, and then transported to fields where they are spread according to nutrient needs. In various places, composting has resulted in improved decomposition rates. Although the use of such alternative organic matter, primarily household waste, has been increasing, overall quantities are rarely sufficient to fertilize entire fields in a sustainable manner.

Maximizing the Efficiency of Nutrient Uptake

Reduced Land Tillage: Although some farmers appreciate tillage for weed control and soil aeration, there seems to be a growing recognition that tillage also destroys the protective vegetative cover and, as a result, exposes soil nutrients to the elements. In areas where ploughs and draught animals are available to smallholders, tillage is still widely practised. In other areas, such as the Senegalese Peanut Basin, farmers have replaced deep tillage by superficial tillage, primarily because of the lack of machinery (Tschakert and Tappan, 2004). For the purpose of CS, reduced or no-tillage are preferable, simply because they enhance the storage of C in the soil.

***Precision Agriculture*:** Many farmers often match crops and management practices with the fertility status of specific fields and, on a smaller scale, with specific spots within one field. Given the relative scarcity of organic and inorganic inputs, the available quantities are spread following a patch-by-patch scheme.

Fire Management

Fire is a very common tool for land managers in drylands. Fire is often used to clear the fields of weeds before planting. Another important reason for preplanting fire is to kill a range of agricultural pests. The role of fire in the soil carbon balance was investigated by modelling and found to have a significant effect on SOC. When the fire-return period was increased from 3 to 15 years, the SOC level increased by 30 percent (Poussart and Ardö, 2002).

These descriptions of individual methods of soil fertility management do not capture the full complexity of the ways in which they are combined. Some of this complexity is described in more detail below.

Building on Local Knowledge

In addition to the practices, it is important to understand farmers' theories of soil fertility, soil formation, and the processes that cause losses and gains of soil fertility over time.

For example, smallholders in the Senegalese Peanut Basin perceive soil fertility as "saletés" (dirt), a generic term for organic matter inputs (manure, decomposing plant material, household waste, etc). To farmers, this "dirt" contains nutritious elements, referred to as "vitamins" or "tasty ingredients" that determine the strength and health of a soil. Although the majority of farmers lack detailed knowledge with respect to the origin of such "vitamins", they are aware of the various processes resulting in soil degradation and fertility losses. The most frequently cited causes of fertility decline include: continuous cultivation without external inputs or crop rotation; reduction of protective vegetation cover; and exposure of SOM to the elements as a consequence of tree removal, deep tillage, and too short fallow periods; bush fires; and harmful insects. Accordingly, farmers' preferred options to restore soil fertility focus on crop rotation, increased organic matter inputs, and accumulation of vegetative cover, primarily through longer-term fallowing and increased tree density.

In parts of Niger, soil fertility is seen much more holistically by farmers than by agronomists, who disaggregate the influences on crop productivity into factors such as water supply, water uptake, individual nutrients, and soil structure (Osbahr and Allen, 2002). The farmers know that the productivity of different soils is determined by a combination of factors.

In a wet year, clay-rich soils in depressions may be waterlogged and unproductive, while sandy soils, where managed adequately, yield at acceptable levels. Clay-rich soils on better-drained sites may be very productive, and very responsive to inputs of manure or fertilizer. In a dry year, the sandy soils are barren, the well-drained clay-rich soils are too hard to till, and only the clay-rich soils of the depressions yield anything at all. Hard-won experience of each subtly different village environment yields a huge variety of different "knowledge" and practices, and a soil CS project could only succeed if all this experience were tapped.

Using farmers' knowledge and practices as an entry point for CS activities offers several advantages: (i) it stimulates farmers' participation in research and project design from the outset; (ii) it facilitates the introduction of C, N and other minerals unknown to the majority of smallholders in a way that is readily understandable and easy to integrate in their own construct of soil theories; and (iii) it opens new doors for extension services to engage, together with farmers, in a more participatory and holistic approach to problem solving instead of delivering predefined agronomic packages.

Realizing the Biophysical Potential for Carbon Sequestration in Farming Systems

In dryland environments, SOC in the first 100 cm soil amounts to about 4 tonnes/ha (Batjes, 1999). This is considerably lower than in other environments. Batjes' estimates for current SOC are: 7 - 10 tonnes/ha in the tropics; 7 - 13 tonnes/ha in the subtropics; 11 - 13 tonnes/ha in temperate regions; and 21 - 24 tonnes/ha in boreal, polar and alpine areas.

Few reliable numbers exist for the entire Sahel, with the exception of estimates for semi-arid savannahs and dry forests in Senegal (the West-Central agricultural region) as reported by Tiessen and Feller, 1998, Ringius, 2002 and Tschakert, Khouma and Senè, 2004, ranging from 4.5 tonnes C/ha for continuously cultivated areas without manure input to 18 tonnes C/ha for non-degraded savannahs (top 20 cm soil).

There is a suite of recommended practices and land-use types that are recommended to increase both the uptake of C from the atmosphere and the duration of storage in soils. As for croplands, FAO (2001b) differentiates practices that decrease carbon losses from the soil from those that increase organic matter inputs into the soil, and considers a combination of both.

The first category includes reduced/conservation/zero tillage, crop residue management, green manuring, cover crops, and integrated weed control.

The second category is based on increases in biomass resulting from manure, compost, mulch farming, mineral fertilization and irrigation as well as improved crop-residue management and green manuring with leguminous species.

All these practices simultaneously increase CS, improve soil fertility, and decrease erosion through soil restoration in drylands, thus offering real potential for a win - win situation for local smallholders.

However, reliable dryland estimates on how much C could be sequestered under the various management practices and farming patterns are still sparse. The most comprehensive estimates range from 0.05 to 0.3 tonnes C/ha/year for croplands and from 0.05 to 0.1 tonnes C/ha/year for grasslands and pastures (Lal *et al.*, 1998). The estimates by Lal *et al.* for tropical areas are about twice as high as those for drylands. For the Old Peanut Basin in Senegal, Tschakert, Khouma and Senè (2004) report a possible range of 0.02 - 0.43 tonnes C/ha/year for improved crop-fallow systems.

Table: *Effects from land management practices or land use on carbon sequestration potential in drylands*

Technological options	***Sequestration potential (tonnes C/ha/year)***
Croplands	
Conservation tillage	0.10 - 0.20
Mulch farming (4 - 6 Mg/ha/year)	0.05 - 0.10
Compost (20 Mg/ha/year)	0.10 - 0.20
Elimination of bare fallow	0.05 - 0.10
Integrated nutrient management	0.10 - 0.20
Restoration of eroded soils	0.10 - 0.20
Restoration of salt-effected soils	0.05 - 0.10
Agricultural intensification	0.10 - 0.20
Water conservation and management	0.10 - 0.30
Afforestation	0.05 - 0.10
Grassland and pastures	0.05 - 0.10

Source: After Lal *et al.* (1998)

Dryland farming systems with access to adequate water resources may benefit from developing their irrigation potential. In small-scale irrigation systems, a high potential for CS arises from four characteristics: from four characteristics:

- The supply of water allows high primary productivity, more so in some (qanat irrigation) than in others (runoff agriculture).
- In most of these systems, the soils are fine-textured, permitting high quantities and slow decay of soil C, much of which is bound closely to clay particles.
- The extensive use of manure, both from animal and crop residues.
- The generally carbon-neutral energy sources, such as animal and human power.

In qanats, the large investment of human power in the early years may be drawn on for centuries, or even millennia.

Larger-scale irrigation schemes should not be dismissed outright as systems for soil CS. Some date back to times when they were no inputs of fossil energy. Therefore, their construction might be regarded as carbon-neutral. Even more recent systems, as in the vast systems in Pakistan and India, were constructed largely with animal or human power. With their high productivity, use of many fine-textured soils and sometimes large extent, these may be potentially valuable sinks of soil C. However, the total carbon budget of each, from the time of construction through to recent interventions, such as deep drains or tubewells, would have to be evaluated separately before it could be considered as suitable as a soil carbon sink. Some would almost certainly not qualify. These would be those in which large amounts of fossil energy had been used in construction, or in which large quantities of fertilizer were being used.

4

Biophysical Aspects of Carbon Sequestration in Drylands

The process of CS, or flux of C into soils forms part of the global carbon cycle. Movement of C between the soil and the atmosphere is bidirectional. Consequently, carbon storage in soils reflects the balance between the opposing processes of accumulation and loss. This reservoir of soil C is truly dynamic. Not only is C continually entering and leaving the soil, the soil C itself is partitioned between several pools, the residence times of which span several orders of magnitude. Nor is soil C an inert reservoir, the organic matter with which it is associated is vital for maintaining soil fertility and it plays a part in such varied phenomena as nutrient cycling and gaseous emissions. A detailed description and analysis of soil C and organic matter can be found elsewhere (Schnitzer, 1991; FAO, 2001c). Taking into account specific biophysical characteristics of dry areas.

Halophytes

A special feature of many dryland soils is salinity, either through natural occurrence or increasingly as a result of irrigation. Saline soils affect large parts of the drylands (Glenn *et al.*, 1993). Such lands are often abandoned, but halophytic plants are especially adapted to these conditions and offer potential for sequestering C in this inhospitable environment. It has been estimated that 130 million ha are suitable for growing halophytes, which can be used for forage, feed and oilseed. Glenn *et al.* estimate that 0.6 - 1.2 gigatonnes of C per year could be assimilated annually by halophytes. Evidence from decomposition experiments suggests that 30 - 50 percent of this C might enter long-term storage in soil. Although irrigation would be required to achieve these figures, a complete carbon budget still suggests a carbon actual rate of 22 - 30 percent.

Grasslands

Grasslands are the natural biome in many drylands, partly because rainfall is insufficient to support trees, and partly because of prevailing livestock management. However, grassland productivity and CS have been controversial. The productivity of tropical grasslands is now known to be much higher than was previously thought, and consequently they sequester much more C (Scurlock and Hall, 1998). Estimates for C stored under grassland are about 70 tonnes/ha, which is comparable with values for forest soils. Although many of the grassland areas in drylands are poorly managed and degraded, they offer potential for CS as a consequence.

The average annual input of organic matter into grassland is about double the 1 - 2 tonnes/ha that is contributed to cropped soils (Jenkinson and Rayner, 1977). This fact is borne out by the results of studies from various locations. The data have shown that grassland, even where subject to controlled grazing, generally has higher soil C levels than cropland. Chan and Bowman (1995) found that 50 years of cropping soils in semiarid New South Wales, Australia, had on average reduced soil C by 32 percent relative to pasture. The reduction was linearly related to the number of years of cropping. Similarly, soils of tall-grass pasture under controlled grazing had greater soil C than adjacent cropland subject to conservation tillage (Franzluebbers *et al.*, 2000).

The key factor responsible for enhanced carbon storage in grassland sites is the high carbon input derived from plant roots. It is this high root production that provides the potential to increase SOM in pastures and vegetated fallows compared with cropped systems. Root debris tends to be less decomposable than shoot material because of its higher lignin content (Woomer *et al.*, 1994). Consequently, the key to maintaining and increasing CS in grassland systems is to maximize grass productivity and root inputs (Trumbmore *et al.*, 1995). Grasses have also been shown to sequester more C than leguminous cover crops (Lal, Hassan and Dumanski, 1999). Grasses also have the potential to sequester C on previously degraded land. Garten and Wullschleger (2000) used a modelling approach that estimated a 12-percent increase in soil C could be obtained under switchgrass (*Panicum virgatum* L.) on degraded land in ten years.

Grazing is a feature of many grasslands, natural or managed. This might be expected to decrease the availability of residues that can be used to sequester C, especially as the quantity of C returned in manure is less than that consumed. However, provided there is careful grazing management, many investigations have found a positive effect of grazing on the stock of soil C. This was found to be the case for a composite pasture (alfalfa and perennial grasses) in the semi-arid pampas (Diaz-Zorita, Duarte and Grove, 2002). Even under harsher conditions in the Syrian Arab Republic, grazing was found to have no detrimental impact on soil C (Jenkinson *et al.*, 1999). Schuman, Janzen and Herrick, (2002) have calculated that with proper grazing management, rangelands in the United States of America can increase soil carbon storage by 0.1 - 0.3 tonnes/ha/ year. For new grasslands, this can rise to 0.6 tonnes/ha/year.

The positive effect of grazing appears to result from the effect that it has on species composition and litter accumulation. Willms *et al.* (2002) found that when prairie was protected from grazing there was little effect on production but there was an increase in the quantity litter. Reeder and Schuman (2002) also found that there was an accumulation of litter in an ungrazed semi-arid system and that soil carbon levels were higher in the grazed lands. The litter acted as a store of immobilized C. The ungrazed grassland also experienced an increase in species that lacked the fibrous rooting system that is conducive to SOM formation and accumulation.

Therefore, grasslands can play a vital role in sequestering C. However, careful grazing management is essential. The historical record shows how susceptible semiarid grasslands are to overgrazing, soil degradation and carbon loss.

Burning

Fires form part of the natural cycle in many biomes and are especially prevalent in grassland ecosystems. However, humans have also used fire to clear areas for agriculture and to clear crop residues. The action of fire would seem to be counter to CS as it returns C fixed by vegetation directly to the atmosphere, thereby preventing its incorporation into the soil. The effect of fire is difficult to generalize because it depends upon the intensity and speed of the fire. These factors are influenced by: the state of the vegetation, i.e. its maturity and woody component; the accumulation of litter; and climate factors such as moisture level. They C in all the aboveground material that is fully combusted will be lost from the system. However, in grassland ecosystems, C lost to fires can be replaced quickly by increased photosynthesis and vegetative growth (Knapp, 1985; Svejcar and Browning, 1988). Even in savannah systems that contain woody species, it has been shown that C lost through combustion can be replaced during the following growing season (Ansley *et al.*, 2002). Regarding the soil, the intensity and speed of the fire will govern the depth to which it is affected. In one study where burning was used to clear forests, 4 tonnes C/ha was lost in the top 3 cm of soil, but this was replaced within one year under a pasture system (Chone *et al.*, 1991).

Not all plant material is fully combusted by fire, and a variable amount of charcoal will be produced. Charcoal is extremely resistant to decomposition; it is not cycled like most organic matter, and has a mean residence time of 10 000 years (Swift, 2001). Consequently, in severely degraded soils that have been affected by several fires, charcoal and charred material can form a substantial proportion of the remaining organic C. However, it is not known whether charcoal and other charred material have any protective effect on the native organic matter. Thus, although fires release CO_2 back into the atmosphere, the simultaneous production of charcoal can be regarded as a sequestration process that results in a substantial amount of C being accumulated within the soil for a long period.

Afforestation

Forestry is recognized as major sink for C. However, as well as accumulating C aboveground, it can also make significant contributions to soil C even in drylands. A number of suitable species are available that enable viable afforestation in dryland environments (Srivastava *et al.*, 1993; Silver, Ostertag and Lugo, 2000; Kumar *et al.*, 2001; Niles, *et al.*, 2002). In particular, N-fixing trees generally lead to increased accumulation of soil C. For example, *Prosopis* and *Acacia* are adapted to subtropical semi-arid lands and are reported to increase the level of soil C by 2 tonnes/ha (Geesing, Felker and Bingham, 2000).

Some tree varieties are particularly suitable for growing on degraded lands, their deep rooting systems tapping resources unavailable to shallow-rooted crops. *Prosopis juliflora* has been grown on salt-affected soils in northwest India and increased the SOC pool from 10 tonnes/ha to 45 tonnes/ha in an eight-year period (Garg, 1998). Even where large-scale forestry is not appropriate, there is often scope for introducing trees around farmers' fields. Such is the case in semi-arid India where *Prosopis cineraria* has improved soil fertility as well as sequestering extra C (Nagarajan and

Sundaramoorthy, 2000). However, natural systems are complex and trees do not guarantee improved CS. Jackson *et al.* (2002) found that soil C decreased when woody vegetation invaded grasslands. Although there was an increase in aboveground and belowground biomass, these gains were offset by losses in soil C.

Residues

Plant residues provide a renewable resource for incorporation into the SOM. Production of plant residues in an ecosystem at steady-state will be balanced by the return of dead plant material to the soil. In a native prairie, about 40 percent of plant production is accumulated in the SOM (Batjes and Sombroek, 1997). However, in agricultural systems, because plants are harvested, only about 20 percent of production will on average be accumulated into the soil organic fraction. Furthermore, in some farming systems, all aboveground production may be harvested, leaving only the root biomass. Of the plant residue returned to the soil, about 15 percent will be converted to passive SOC (Lal, 1997). Schlesinger (1990) is more pessimistic, suggesting that only 1 percent of plant production will contribute to CS in soil. The actual quantities of residue returned to the soil will depend on the crop, the growing conditions and the agricultural practices. For example, for a soybean - wheat system in subtropical central India, the annual contribution of C from aboveground biomass was about 22 percent for soybean and 32 percent for wheat (Kundu *et al.*, 2001). This resulted in 18 percent of the annual gross carbon input being incorporated into the SOM. In semi-arid Canada, the conversion of residue C to SOC was reported to be 9 percent in frequently fallowed systems, increasing to 29 percent for continuously cropped systems (Campbell *et al.*, 2000).

Unless a root crop is being harvested, all belowground production is available for incorporation into the SOM. Roots are believed to be the major constituent of particulate organic matter although tillage reduces the net accumulation of C from roots substantially (Hussain, Olsson and Ebelhar, 1999). In cool climates, belowground carbon inputs from roots alone can generally maintain soil carbon levels. However, this is not the case in warmer or semi-arid regions where residues are decomposed much more readily, providing sufficient moisture is available (Rasmussen, Albrecht and Smiley, 1998). Consequently, when continuous cropping is practised in drylands, failure to return aboveground plant residue will lead invariably to a reduction in soil C. Many African soils demonstrate this phenomenon. Continuous cropping over a number of years without the recommended inputs has often halved their carbon content (Woomer *et al.*, 1997; Ringius, 2002).

Both the quality and quantity of plant residues are important factors for determining the amount of C stored in soil. The quantity is highly dependent on the environmental conditions and agricultural practices. Differences between crops can be marked. A crop of maize will return nearly twice as much residue to the soil compared with soybean and, consequently, will result in a higher rate of SOM increase (Reicosky, 1997). The advantage that cereals have over legumes for achieving maximum CS rates has also been demonstrated by Curtin *et al.* (2000). They have shown that while black lentil fallow in semi-arid Canada added between 1.4 and 1.8 tonnes C/ha, a wheat crop would add 2 - 3 times this amount of C annually. Similarly, in Argentina, soybean, which produced 1.2 tonnes/ha of residue, resulted in a net loss of soil C, while maize, with 3.0 tonnes/ ha of residue, lessened the loss of soil C from the system significantly (Studdert and Echeverria, 2000).

Even within one crop group, large differences in organic matter production occur. Abdurahman *et al.* (1998) compared dry leaf production from pigeon pea and cowpea. While the former yielded 3 tonnes/ha, cowpea produced 0.14 tonnes/ha. These examples illustrate how the choice of crop can have a major influence on how much C an agricultural system can sequester.

The importance of roots relative to shoots for providing soil C is another factor illustrated in an experiment to compare the fate of shoot- and root-derived C (Puget and Drinkwater, 2001). In this study with leguminous green manure (hairy vetch), nearly half of the root-derived C was still present in the soil after one growing season whereas only 13 percent of shoot-derived C remained. This implies that shoot residues are broken down rapidly on account of their higher N content (Woomer *et al.*, 1994) and may serve as a nitrogen source for the following crop.

The chemical composition of plant residues affects their rate of decomposition. On average, crop residues contain about 40 - 50 percent of C but N is a much more variable component. A high concentration of lignin and other structural carbohydrates together with a high C:N ratio will decrease the rate of decomposition. For example, measurement of CO_2 evolution from tree leaves of African browse species and goat manure showed a significant correlation with initial nitrogen content and a negative correlation with lignin content (Mafongoya, Barak and

Reed, 2000). Legume residues such as soybean are generally of high quality (low C:N ratio) and so decompose rapidly (Woomer *et al.*, 1994). Although the chemical composition of the plant residue affects its rate of decomposition, there is little effect on the resulting SOM (Gregorich *et al.*, 1998).

Where residues accumulate on the soil surface, their physical presence affects the soil. Mulches reduce water loss and soil temperature (Duiker and Lal, 2000), both important factors for drylands, especially where the soil temperature is above the optimum for plant growth. The ability of soil to assimilate organic matter is not clear-cut. A linear relationship between application and accumulation of SOM is often quoted. However, measurement of CO_2 flux in central Ohio, the United States of America, showed this flux to increase with added application of wheat straw (0, 8 and 16 tonnes/ha). However, the SOC was 19.6, 25.6 and 26.5 tonnes/ha after four years, suggesting that CS was reaching saturation (Jacinthe, Lal and Kimble, 2002).

Care must be taken when applying residues, as large losses of C can still occur under certain conditions. For example, in western Kenya, 70 - 90 percent of the added C was lost within 40 d when green manure from agroforestry trees was applied during the rainy season (Nyberg *et al.*, 2002). In Niger, the addition of millet residue and fertilizer for five years had no significant effect on the carbon level in the sandy soils (Geiger, Manu and Bationo, 1992). Termite activity may have contributed to the low levels of soil C here because entire surface mulches can be consumed within one year.

The potential amount of residues available for applying to soils can be large. Gaur (1992) estimated that in India about 235 million tonnes of straw/stover are produced annually from five major cereals (not only from drylands). Even if half of this were used for feeding livestock, there would be more than 1 million tonnes available for adding to the soil. However, availability of sufficient plant residues is often a problem, especially where they are required for livestock feed. This conflict of requirement occurs frequently in many dryland farming systems. In West Africa, crop residues are either removed or burnt. Consequently, the amount of soil C has declined steadily. Continuous cultivation and manure application can raise soil carbon levels by 40 percent but this frequently involves mining C from neighbouring areas to support the livestock (Ringius, 2002). Where plant and animal residues are in short supply, possibilities may exist for alternative organic inputs to the soil. For example, in India, waste products from a plant processing coir dust have been incorporated successfully into soil (Selvaraju *et al.*, 1999), and other experiments with industrial glue waste (Dahiya, Malik and Jhorar, 2001) have increased soil carbon levels successfully.

With regard to the complete carbon budget, where plant residues accumulate *in situ*, there is no extra carbon cost involved. Consequently, the C fixed by plants in photosynthesis is available as a net gain to the soil. The situation becomes more complex where machinery is required to separate the residue from the harvestable components. Where plant residues are transported between fields the energy cost would need to be included.

Perhaps the major issue regarding residue application and carbon accounting is the question of whether C is simply being transferred from one place to another rather than being truly sequestered. If organic material from industrial processes or other sources were to be used for incorporation into the soil, then any C used in transport would have to be accounted for. However, where the material is truly a waste product, there should be no need to consider the C used in its production. In this case, the C available for sequestration should be viewed no differently from the CO_2 emitted from a fossil fuel source that is subsequently fixed by plants in photosynthesis and returned to the soil via crop residues.

Applied Manures

The application of FYM has long been treated as a valuable source of organic matter to enhance soil fertility. One of the key characteristics of manure application is that it promotes the formation and stabilization of soil macroaggregates (Whalen and Chang, 2002) and particulate organic matter (Kapkiyai *et al.*, 1999). Manure is more resistant to microbial decomposition than plant residues are. Consequently, for the same carbon input, carbon storage is higher with manure application than with plant residues (Jenkinson, 1990; Feng and Li, 2001). Following five years of application, soil receiving manure had 1.18 tonnes/ha more C present than soil receiving plant residues. Even after 15 years, there was a still a difference of 0.37 tonnes C/ha as calculated by the RothC soil carbon model. In the field, Gregorich *et al.* (1998) found that manured soils had large quantities of soluble C with a slower turnover rate than in control or fertilized plots.

The composition and, therefore, decomposition of manure varies between the species from which it originates and also within species according to their diet (Somda and Powell, 1998). Many field trials have found that manure is the best means for incorporating organic matter into soils and promoting carbon storage. For example, Li *et al.* (1994) found that manure yielded the largest amount of C sequestered over a range of soils and climate conditions, although soil texture was important, and the greatest rate of sequestration occurred where there was a high clay content. However, many traditional agrosystems add manure in combination with fertilizers (Haynes and Naidu, 1998).

Depending on the system, the application of even relatively high amounts of FYM does not guarantee an increase in soil C. A long-term study in Kenya has shown that SOM declined even when manure was applied and maize residue returned (Kapkiyai *et al.*, 1999). It has been estimated that in order to maintain soil C in this system, 35 tonnes/ha of manure or 17 tonnes/ha manure with 16 tonnes/ha of stover would be required annually (Woomer *et al.*, 1997). Consequently, there is a carbon shortfall in this case, but it is difficult to see how the present system could remedy it. In addition, high application rates of manure can sometimes cause problems in the soil through the accumulation of K^+, Na^+ and NH_4^+ and the production of water-repellent substances by decomposer fungi (Haynes and Naidu, 1998). An additional problem in drylands that restricts the quantity of manure that can be applied is "burning" of the crop when insufficient moisture is available at the time of application. Consequently, farmers often wait until the rains have come before making an application, especially as precipitation is often erratic in arid regions. is often erratic in arid regions.

The production of sufficient manure for application to fields is a real problem for many smallholder farming systems. In Nigeria, direct manure input onto land from dry-season grazing is about 111 kg/ha of dry matter (Powell, 1986). This quantity will have little effect on the soil. More useful is the practice of night-parking cattle as manure production is usually greatest at dawn and dusk. When 50 cattle were penned in an area of 0.04 ha for five nights, they produced the equivalent of 6.875 tonnes/ha of manure (Harris, 2000). Normally, cattle will be penned in fields for 2 - 3 nights in northern Nigeria and this can supply manure at a rate of 5.5 tonnes/ha. Alternatively, in densely populated parts such as the Kano closed-settlement zone, cattle and crop production are fully integrated. Cattle are kept permanently in pens and fed using feed grown on neighbouring fields. Their manure is collected and spread onto the croplands. Although an efficient system, some C will be lost as a consequence of the respiratory and growth requirements of the cattle. An additional problem associated with cattle rearing is that ruminants produce significant quantities of CH_4, which is a potent GHG.

There has been some debate as to the usefulness of animal manure in CS. Schlesinger (1999, 2000) has calculated that providing 13.4 tonnes/ha would require 3 ha of cropland in order to produce sufficient cattle feed. This means that manure production requires mining of C from neighbouring lands.

Although this is a generalization, it is argued that the 3:1 differential makes it unlikely that manure production per se could be used as a means of providing a net carbon sink in soils. However, in many dryland smallholder cropping systems, manure application rates are much lower, and fodder production is also less efficient.

Smith and Powlson (2000) have made the case that keeping cattle is part of many agricultural systems and, hence, that manure should be considered as a byproduct that can be added to arable land without necessarily including the carbon cost of its production.

Part of the disagreement with Schlesinger on the usefulness of manure depends on where the system boundary for carbon accounting is drawn. However, when conducting such a carbon audit, it is essential to remember that the purpose of agriculture is to feed people; offsetting GHGs can only be evaluated as a secondary activity.

Inorganic Fertilizers and Irrigation

Fertilization and irrigation are primary means for increasing plant production and crop yield. Any increase in biomass also offers increased scope for CS. Consequently, irrigation and fertilization have been recommended as, and proved to be, successful methods of increasing CS (Lal, Hassan and Dumanski, 1999). Rasmussen and Rohde (1988) have shown a direct linear relationship between long-term nitrogen addition and accumulation of organic C in some semi-arid soils in Oregon, the United States of America. However, these technologies provide no additional organic matter themselves but do carry a carbon cost. Schlesinger (1999, 2000) has pointed out that pumping water requires energy and that the process of fertilizer manufacture, storage and transport is energy intensive. Consequently, Schlesinger (2000) has estimated that the gains in C stored using either fertilization or irrigation are offset by losses elsewhere in the system. Irrigation can also lead to the release of inorganic C from the soil.

Izaurralde, McGill and Rosenberg (2000) have argued that the calculations used by Schlesinger are based on very high rates of fertilizer application. For many dryland agriculture systems in the developing world, farmers do not have sufficient funds to apply large quantities of fertilizer even if they were available. With regard to the energy costs incurred in pumping water, solar-powered systems are being developed (Sinha *et al.*, 2002) and dryland environments with frequently clear skies would be able to make the best use of solar radiation.

These examples serve to illustrate how important it is to consider the whole system when CS is being considered to offset CO_2 emissions. The exact carbon cost of irrigation and fertilization would require calculation for each system, but the carbon deficits associated with both technologies make their incorporation into net CS systems difficult to achieve. Water conservation, the growing of legumes and careful nutrient cycling are more likely to yield a positive carbon balance.

Tillage

Pretty *et al.* (2002) consider tillage to be one of the major factors responsible for decreasing carbon stocks in agricultural soils. Research and experimentation with reduced tillage practices are most prevalent in the Americas. The mould-board plough and disc harrow are believed to be the causes of the loss of soil C through the destruction of soil aggregates and the acceleration of decomposition by the mixing of plant residues, oxygen and microbial biomass. Soil aggregates are vital for CS (Six, Elliott and Paustian, 2000), a process that is maximal at intermediate aggregate turnover (Plante and McGill, 2002). Of the organic matter fraction, the particulate organic matter is the most tillage sensitive (Hussain, Olsson and Ebelhar, 1999).

It is difficult to quantify the effects of tillage on soil C because the effect are very site dependent, e.g. coarse-textured soils are likely to be more affected by cultivation than are fine ones (Buschiazzo *et al.*, 2001). However, reducing tillage should be most effective in hot, dry environments (Batjes and Sombroek, 1997).

Reicosky (1997) conducted an experiment that used measurements of CO_2 efflux to investigate tillage-induced carbon loss from soil. The flux of CO_2 was monitored for 19 d following different forms of tillage practice. The mould-board plough buried most of the crop residue and produced the maximum CO_2 flux. The C released by the different treatments as a percentage of C in the crop residue was: 134 percent with mould-board plough; 70 percent with mould-board plough and disc harrow; 58 percent with disc harrow; 54 percent with chisel plough; and 27 percent with no-tillage. This demonstrates the correlation between CO_2 loss and tillage intensity, and demonstrates why farming systems that use mould-board ploughing inevitably lose soil C. Very large amounts of organic matter would be required to replace the loss incurred by such heavy tillage. Reicosky *et al.* (1995) estimate that 15 - 25 tonnes/ha manure plus crop residue would be needed annually in North America to offset these losses.

The flux of CO_2 from soil generated directly by the tillage process may not always reflect the overall release of CO_2 and hence carbon storage of the system. This is illustrated by a comparison of conventional disc tillage and no-tillage in central Texas, the United States of America, by Franzluebbers, Hons and Zuberer (1995). Here, seasonal evolution of CO_2 was up to 12 percent greater in the no-tillage system after 10 years. This was despite the fact that more C was sequestered by the no-tillage system. The authors suggest that a change in the dynamics of CS and mineralization have occurred under the no-tillage system. Similarly, Costantini, Cosentino and Segat (1996) found that more CO_2 was released from no-tillage or reduced-tillage compared with conventional tillage despite there being increased levels of soil C. They ascribe this difference to an increase in the microbial biomass.

Rates of carbon loss through tillage depend considerably on the site and cropping system. Ellert and Janzen (1999) measured the flux of CO_2 following the passage of a heavy cultivator on a semi-arid Chernozem soil in the Canadian prairies. They found that although tillage increased rates of CO_2 loss by two to four times, values returned to normal after 24 hours. They calculated that even with ten passes of the cultivator, only 5 percent of crop residue production would be released from this cropping system. In another situation, ploughing of a wheat - fallow cropping system near Sydney, Australia, reduced soil C by 32 percent after 12 years. Elimination of ploughing and adopting a no-tillage approach was unable to prevent a decrease in the carbon stock, although the loss was reduced to just 12 percent (Doran, Elliott and Paustian, 1998). The authors suggest that a fallow period would be required to halt the decline in soil C at this site.

There are many different types of tillage system. Conservation tillage covers a range of practices - no-tillage, ridge-tillage, mulch-tillage (Unger, 1990). Mulch-tillage maintains higher levels of residue cover. With mulches, only a small fraction of the residue is in contact with the soil surface and the microbes it contains. Decomposition is slow, especially as oxygen availability is limited. The physical presence of crop residues on the soil surface also alters the microclimate of the upper soil layer, which tends to be cooler and wetter compared with conventional tillage (Doran, Elliott and Paustian, 1998).

The accumulation of residues also reduces the loss of CO_2 from the soil surface. Alvarez *et al.*, (1995) reported an increase in labile forms of organic matter under no-tillage in the Argentine rolling pampa, indicating a decrease in the mineralization of the organic fraction. This study also noted that although organic C increased by 42 - 50 percent under no-tillage compared with ploughing and chisel tillage, there was also a marked stratification in the distribution of C under the no-tillage regime that was not evident in the ploughed system.

Stratification of organic C is common with reduced or no-tillage. Zibilske, Bradford and Smart (2002) in semi-arid Texas, the United States of America, demonstrated that the organic carbon concentration was 50 percent greater in the top 4 cm of soil of a no-tillage experiment compared with ploughing, but the difference dropped to just 15 percent in the 4 - 8-cm depth zone. This is typical of organic carbon gains observed with conservation tillage in hot climates. Bayer *et al.* (2000), working on a sandy clay loam Acrisol, also found that the increase in total organic C was restricted to the soil surface layers under no-tillage but that the actual quantity depended on the cropping system. An oat/vetch - maize/cowpea no-tillage system produced the largest quantity of crop residues and sequestered the most C: 1.33 tonnes C/ha/year in 9 years.

Reicosky (1997) has compared the results from many no-tillage trials. The data emphasize the effect that crop rotation and quantity of crop residue has on organic matter accumulation. Overall, rates of organic matter accumulation can be expected to be lower in the hotter climates. Nevertheless, even in the very sandy soils of in north of the Syrian Arab Republic, it has been possible to make modest increases in SOM with no-tillage (Ryan, 1997). In western Nigeria, no-tillage combined with mulch application had a dramatic effect, increasing soil C from 15 to 32.3 tonnes/ha in 4 years (Ringius, 2002).

Although no-tillage systems are an excellent tool for combating the carbon losses associated with conventional cultivation, they do have their own special problems. In temperate lands, the reduction in soil temperature commonly associated with plant residue accumulation on the soil surface can retard germination. However, in drylands, where soil temperatures are frequently above the optimum for germination and plant establishment, such cooling is likely to be beneficial (Phillips *et al.*, 1980). No-tillage systems frequently suffer from an increased incidence of pests and diseases; the mouldboard plough and disc harrow are efficient weed controllers (Reicosky *et al.*, 1995). Consequently, no-tillage systems generally rely upon extra herbicides and pesticides. These inputs have an economic price and they also incur a carbon cost. However, in many dryland farming systems of the developing world, purchasing such products is not feasible, and quite often there is plentiful labour available for weeding. Application of N fertilizer can also be problematic when used on undisturbed, no-tillage soils. Where the soil is poorly drained, denitrification can occur and the reduced rate of evaporation increases the risk of nitrate leaching. In addition the native soil N has a lower rate of mineralization in undisturbed soil.

Not all soils are suited to a reduced-tillage approach. Some soils in the Argentine pampa may actually lose more C under no-tillage (0.7 - 1.5 tonnes C/ha/year) compared with conventional ploughing (Alvarez *et al.*, 1995) and periodic ploughing is required to avoid soil compaction (Taboada *et al.*, 1998). Where no-tillage is used on the pampas, the physical status of the soil is a critical factor for the success of the system (Diaz-Zorita, Duarte and Grove, 2002). Similarly, in the West African Sahel, the highest crop yields are obtained with deep ploughing, which is required to prevent crusting and alleviate compaction. In general, the success of reduced-tillage systems is often dependent on soil texture (Needelman *et al.*, 1999).

A particular advantage of the no-tillage system is that it favours multicropping; harvesting can be followed immediately by planting (Phillips *et al.*, 1980). Any cropping system that allows for continuous or near continuous plant growth should yield the maximum capacity to produce plant biomass and, consequently, has the potential to provide the greatest amount of organic mater for inclusion into the soil.

Considering the overall carbon budget, no-tillage systems have a lower energy requirement because tillage is very energy intensive. Phillips *et al.* (1980) have calculated that no-tillage systems in North America reduce the energy input into maize and soybean production by 7 and 18 percent, respectively. Improved water-use efficiency means that the energy, and hence carbon, cost of irrigation are reduced. However, the impact of energy savings is frequently offset by additional herbicide requirements (Phillips *et al.*, 1980). Kern and Johnson (1993) estimated that the manufacture and application of herbicides to no-tillage systems of the Great Plains is equivalent to 0.02 tonnes C/ha.

Reduced-tillage systems were adopted originally to help combat soil degradation. They were not intended as a means of sequestering C, which is a fortunate side-effect. Although the effectiveness of no-tillage at sequestering C will depend on the specific agricultural system to which it is introduced, there is no doubt that, as the intensity of tillage decreases, the balance between carbon loss and gain swings toward the latter.

Rotations

The importance of rotation in agricultural systems has long been known and the procedure now forms an intricate part of many conservation tillage practices. The inclusion of rotations has many benefits such as countering the buildup of cropspecific pests and, thereby, lessening the need for "carbon costly" pesticides and herbicides. Different crop species have a variety of rooting depths and this aids in distributing organic matter throughout the soil profile. In particular, deep-rooting plants are especially useful for increasing carbon storage at depth, where it should be most secure. The inclusion of N-fixing varieties in a rotation increases soil N without the need for energy-intensive production of N fertilizers.

The beneficial effects that rotations have for CS have been proved in many longterm field experiments. For example, Gregorich, Drury and Baldock (2001) made a comparison of continuous maize cultivation with a legume-based rotation. The rotation had a greater effect on soil C than did fertilizer.

The difference between monocultured maize and the rotation was 20 tonnes C/ha while the effect of fertilization was 6 tonnes C/ha after 35 years. In addition, the SOM present below the ploughed layer in the legume-based rotation appeared to be more biologically resistant. This demonstrates that soils under legume-based rotations tend to preserve residue C. A positive effect on SOC (an increase of 2-4 tonnes/ha) was also found with legumes and alternate cattle grazing in semi-arid Argentina (Miglierina *et al.*, 2000).

Rotations, especially legume-based ones, are generally regarded as extremely valuable for maintaining soil fertility and have a very good potential for sequestering C in dryland systems. Drinkwater, Wagoner and Sarrantonio (1998) estimate that their use in the maize/soybean-growing region of the United States of America would increase soil CS by 0.01 - 0.03 Pg C/year. The effectiveness of rotations for sequestering C is likely to be greatest where they are combined with conservation tillage practices.

Fallows

The role that fallows play in CS is varied. Where cropping is not taking place, it is important that vegetation cover is preserved. This is especially so in drylands where exposed soil is most likely to suffer from erosion and degradation. In addition to protecting the soil, cover crops can utilize solar energy that would otherwise be wasted. The CO_2 fixed is then available for sequestering into the soil as the plants senesce. The importance of vegetation cover can be illustrated with the results from an experiment conducted at a semi-arid site in Mediterranean Spain (Albaladejo *et al.*, 1998). Four and a half years after the vegetation cover was removed from one site, the SOC had decreased by 35 percent compared with the control plots.

The type of fallow is important. In Nigeria, forest clearance caused a decline in soil C from 25 to 13.5 tonnes/ha in seven years, but 12 - 13 years of bush fallow restored the carbon content (Juo *et al.*, 1995). Conversely, pigeon pea fallow was unable to sequester sufficient C on account of its low biomass production and rapid degradation.

However, fallows can have a negative effect on carbon storage in many situations. The frequency of summer fallows in semi-arid regions has been suggested as one of the major factors influencing the level of soil C in agricultural systems (Rasmussen, Albrecht and Smiley, 1998). Reducing the summer fallow in the semi-arid northwest United States of America is reported to have had a more positive effect on soil carbon retention than that achieved by decreasing tillage intensity. The loss of C in this region is believed to reflect the high rates of biological oxidation that occur here, which can only be offset by very large applications of manure (Rasmussen, Albrecht and Smiley, 1998). Consequently, yearly cropping and the associated organic additions is the recommended practice. Miglierina *et al.* (2000) also found that reducing summer fallow increased soil C, a consequence of the additional crop residue that was added. Using the CENTURY agro-ecosystem model, Smith *et al* (2001) predicted that reducing summer fallow in wheat cropping systems (wheat - fallow to wheat - wheat - fallow) in the semiarid Chernozems of western Canada would reduce carbon losses by 0.03 tonnes/ha.

Elimination of fallows can be highly beneficial for soil C simply because most fallows are associated with small inputs of plant residue. The significance of fallows for CS in a given system will depend on whether or not the cropping cycle adds significant quantities of organic matter to the soil. Where it does, then the presence of fallows is unlikely to enhance carbon storage within the system. Conversely, where the cultivation practice is poor and little or no organic matter is added, fallow periods will serve to counter this situation.

Soil Inorganic Carbon

Not all soil C is associated with organic material; there is also an inorganic carbon component in soils. This is of particular relevance to drylands because calcification and the formation of secondary carbonates is an important process in the soils of arid and semi-arid regions where, as a result, the largest accumulations of carbonate occur (Batjes and Sombroek, 1997). The dynamics of the inorganic carbon pool are poorly understood although it is normally quite stable. Sequestration of inorganic C occurs via the movement of HCO^-_3 into groundwater and closed systems. According to Schlesinger (1997) accumulation of calcium carbonate is quite low at 0.0012 - 0.006 tonnes/ha. However, Lal, Hassan and Dumanski (1999) believe that the sequestration of secondary carbonates can contribute 0.0069 - 0.2659 Pg C/year in arid and semi-arid lands.

Although soil inorganic C is relatively stable, it will release CO_2 if the carbonates become exposed through erosion (Lal, Hassan and Dumanski, 1999). In addition, irrigation can cause inorganic C to become unstable if acidification takes place through inputs of N and sulphur. The release of CO_2 through carbonate precipitation is seen as a major problem if irrigation waters are used in any system that is trying to store C. Furthermore, Schlesinger (2000) has pointed out that the groundwater of arid lands often contains up to 1 percent calcium and CO_2. This concentration is much higher than that which occurs in the atmosphere. Consequently, when these waters are applied to arid lands, CO_2 is released to the atmosphere and calcium carbonate precipitates. Schlesinger's calculations suggest that irrigation of some cropping systems would yield a net transfer of CO_2 from the soil to the atmosphere.

Trace Gases

An important aspect of agricultural systems in relation to the global carbon balance is the production of trace gases, particularly CH_4 and N_2O. When CS by soils is being considered as a mechanism for offsetting greenhouse gas emissions, it is necessary to consider all the interacting factors that can influence global warming. Both CH_4 and N_2O are radiatively active gases and, like CO_2, contribute to the greenhouse effect. Although they are present in the atmosphere at much lower concentrations than CO_2, they are much more potent. CH_4 and N_2O are, respectively, 21 times and 300 times more active GHGs than CO_2.

Ruminants, composting, biomass burning and waterlogging produce CH_4, while N_2O is released from soils when N fertilizer or manure is applied (Vanamstel and Swart, 1994). Manure usage is considered to be the major problem with regard to trace-gas emissions in agriculture. This is a potentially serious problem because the application of manure is a major tool for increasing soil C in drylands. Smith *et al* (2001) calculated that, for European soils, the effect of trace gases is sufficient to reduce the CO_2 mitigation potential of some no-tillage and manure-management practices by up to a half. Moreover, climate change is likely to amplify the problem as increased temperature is predicted to promote N_2O emissions (Li, Narayanan and Harriss, 1996).

Climate Change

Climate is a major factor involved in soil formation. Consequently, climate change will influence soils. Photosynthesis and decomposition will be affected directly and, hence, have an impact on soils. Whether soil carbon levels increase or decrease will depend on the balance between primary production and decomposition (Kirschbaum, 1995). Overall, productivity is predicted to increase as a consequence of rising CO_2 concentration and temperature, and this will lead to increased amounts of residue available for incorporation into the soil. However, higher temperatures can be expected to increase mineralization of SOM because this process is more sensitive to temperature increases than primary production. Kirschbaum (1995) predicts that SOC stocks will decline overall with global warming. However, Goldewijk *et al.* (1994) suggest that the effects of temperature and water availability on soil respiration will be smaller than those attributable to the CO_2 fertilization effect. The direction of change is not certain but the balance of change will most probably operate at the regional level.

It has been argued that agricultural systems are to some extent buffered from environmental effects, while decomposition is not protected (Cole *et al.*, 1993). Hence, increased rates of mineralization might be more significant than any enhancement of production. However, the quality of plant organic matter is expected to decrease under elevated CO_2 owing to an increase in the C:N ratio. This would slow the rate of degradation (Batjes and Sombroek, 1997). Globally, the drylands are expected to become moister (Glenn *et al.*, 1993), which should lead to an increase in productivity and decomposition. However, shifts in climate zones are dependent upon a complex array of variables. Predictions based on the CENTURY agro-ecosystem model suggest that, overall, grasslands will lose soil C except in tropical savannahs, which should show a small increase (Parton *et al.*, 1995). Experiments at elevated CO_2 have also shown that changes in soil C in agro-ecosystems are particularly dependent upon the crop species grown (Rice *et al.*, 1994).

The full extent of the global rise in temperature associated with climate change may not be felt in many of the drylands because warming is predicted to be greatest at higher latitudes. With regard to the vegetation, some of the best-adapted plants for dryland regions use the C4 photosynthetic pattern. Because these species already have a CO_2-concentrating mechanism, they show little or no increase in productivity at elevated CO_2. However, they are still likely to receive some benefit from the increased water-use efficiency that accompanies a rising CO_2 concentration.

Case Studies on Drylands

Most of the research and case studies on soil CS have been conducted in temperate zones. Much less work has been done in developing regions, including drylands (Lal, 2002b). In order to realize the great potential for CS offered in drylands, efforts are needed to identify soil-specific practices that restore SOC in degraded soils. This reviews specific case studies in different representative agrosystems in drylands, where the potential for carbon storage is assessed under different land-use and landmanagement practices involving irrigation and biofertilizers.

Models for Analysing Tropical Dryland Agricultural Systems

RothC (RothC-26 3) (Coleman and Jenkinson, 1995; Jenkinson and Rayner, 1977) and CENTURY 4.0 (Parton *et al.*, 1987; Parton, Stewardt and Cole, 1988) are the most extensively used SOC simulation models. They have been tested against a variety of long-term agricultural field trials and have also been used in a variety of climate zones, including dryland regions. Both have also been adopted for use in major carbonassessment projects. The two models vary in their complexity. RothC requires fewer data inputs and so is easier to parametize. However, it only deals with soil processes and, consequently, plant residue C is a required input. The CENTURY agro-ecosystem model has similar SOC pools to RothC but has the advantage of additional submodels. Although it is able to handle a great many more land-management options than RothC, it requires an increased array of input variables that require parametization. This is important because the ability of any model to predict accurately depends on the accuracy and trustworthiness of the data used to parametize it.

Approach Adopted for Parametizing Rothc and Century

Data for parametizing the models can be divided into three areas: climate, soil and land management. Studies that contain data that are sufficiently detailed for modelling are few, particularly in dryland regions. Where investigations have been undertaken and detailed information collected, the primary data that are vital for modelling are often not readily available.

Climate parameters may be referred to by investigators reporting the effects of land management on soils, but complete data sets are rarely given. However, climate data are available independently, as from FAOCLIM 2 (2000), which contains a database of more than 28 000 weather stations worldwide.

The literature contains the results from many investigations analysing soil properties and the effect of various treatments and practices upon soils. However, there are far fewer studies that combine soil analysis with examination of land management, especially over the longer term. In particular, there has been little research on dryland systems.

For the case studies examined here, CENTURY was parametized and run to equilibrium for between 2 000 and 5 000 years. Scenarios reflecting the recent past and present were then applied. Although CENTURY can run many varied cropping practices, it can only handle one crop at a time. Consequently, intercropping, which is commonly practised in dryland farming systems, could not be incorporated into the scenarios.

RothC was run to equilibrium using the current soil carbon status after initially being run in reverse mode to calculate the necessary plant carbon inputs. In order to model some of the future scenarios for analysing the effects of land management on soil C, the appropriate plant residue inputs were obtained from the CENTURY plant submodel and then used to parametize the RothC land-management files. The results in RothC frequently predict higher levels of soil C than CENTURY. This has been noted before and has been attributed to the fact that SOC tends to turnover faster in CENTURY than in

RothC, and, consequently, that RothC requires lower carbon inputs to maintain the same organic carbon content (Falloon and Smith, 2002). The CENTURY plant submodel is quite basic. Therefore, where accurate estimations of plant production and crop yield are required, then alternative models, such as the fully mechanistic plant productivity model WIMOVAC, should be used (Humphries and Long, 1995).

Choice of Systems and Sources of Data

Data from four distinctly different dryland systems in Argentina, India, Kenya and Nigeria were used to model changes in soil C with a variety of farm practices and technologies. These systems had different precultivation stocks of soil C, and had lost different amounts during cultivation. Additions of organic matter to the soil through use of FYM, green manures, legumes in rotations, vermicompost, or use of fallows in rotations, all increased soil C and increased agricultural yields. Trees as part of agroforestry systems further increase soil carbon stocks. Inorganic fertilizer used alone to increase nutrient supply for crops results in declines in soil C in all systems, or only small increases if used with no-tillage. No-tillage increases soil C, although again the accumulation is greatest where organic matter is added to the soil.

The scenarios show that CS in tropical drylands soils can be achieved at the different sites. The land-management practices have been chosen to be in accordance with the current farming systems. Thus, for example, application rates of organic matter are commensurate with quantities that should be available to local farmers. However, at the field level, important trade-offs may occur, preventing adoption of the best strategies for CS. Crop residues may be required for livestock feed or fuel rather than be returned to the fields, or may be sold in difficult times. Animal manures may be burnt for fuel. Many socio-economic factors will interact to determine which scenario or combination of scenarios is implemented in each growing season.

Some of the results predict that soil C can be restored to precultivation levels, or in certain circumstances to above them. The true "native soil carbon level" is often difficult to establish in those systems where agricultural activity has been present for several centuries or millennia such as in the Nigeria and Kenya cases. To achieve quantities of soil C in excess of the "natural level" implies that the agriculture system has a greater productivity than the native system, assuming that C is not being imported. The scenarios that predict the highest CS rates are often associated with the introduction of trees to the system. The inputs of C from trees are more resistant to decomposition than those from herbaceous crops and, consequently, can cause marked increases in the level of soil C (Falloon and Smith, 2002).

Table: *Summary of findings on carbon stocks and rates of accumulation and/or loss in four dryland agrosystems*

	Nigeria	*India*	*Kenya*	*Argentina*
Stocks of soil C before cultivation (tonnes/ha)	8-23	15-20	33-41	50-70
Current stocks of soil C after cultivation (tonnes/ha)	6-12	13-22	18-28	37-41
Effect of conventional tillage practices on soil C (tonnes/ha/year)	- 0.05 to - 0.01	- 0.07 to +0.06	- 0.3 to - 0.1	- 0.17 to - 0.19
Effect of FYM, organic additions, retained plant residues and fallows in rotations (tonnes/ha/year)	+ 0.1 to + 0.3	+ 0.2 to + 0.4	+ 0.4 to + 0.9	-
Effect of trees (tonnes/ha/year)	+ additional 0.05 to 0.15	+ additional 0.5 to 0.7	-	-
Effect of using inorganic fertilizers as sole source of nutrients on soil C (tonnes/ha/year)	- 0.12 to + 0.08	- 0.01	- 0.3	-
Effects of no-tillage (NT)				
NT alone				+ 0.02 +0.1 to 0.25
NT + green manures or FYM				+0.04
NT + inorganic fertilizers				

[1]*Effects of conventional tillage are averages for the last 100 years for each site except Kenya where rates are calculated from when each settlement commenced (30 - 50 years).*

Case Study 1 - Nigeria - Kano Region

Nigeria includes some of the most densely inhabited areas of semi-arid West Africa. This is not a recent phenomenon as indications of human activity can be traced back well over a thousand years. Consequently, the soils of this region have been subjected to long periods of cultivation. However, in the last 40 years the cultivated area of northern Nigeria has increased from 11 percent to 34 percent of the total land area (Harris, 2000). In particular, the level of agricultural intensity in the Kano Closed - Settled Zone (CSZ) is among the highest in semi-arid West Africa.

Investigations into the farming systems of this region and their effect on soil fertility have been conducted by the organization Drylands Research, the United Kingdom. Its studies have found a close correlation between the intensity of farming and the adoption of soil-fertility management techniques. Plant production in the Sahel is generally limited by rainfall or nutrients (Breman and De Wit, 1983). However, the economy and infrastructure of northern Nigeria is unsuited to high external inputs such as fertilizers. Consequently, the smallholder farming units operate as low-input systems, with cattle manure being the most common form of organic input.

Physical Aspects

This semi-arid region has a rainy season from May - July until September. However, rainfall is erratic (Mortimore, 2000) and makes crop cultivation particularly difficult. The soils are principally ferruginous tropical soils that are sandy, with poor waterholding capacity and low levels of nutrients and organic matter (Harris, 2000). Nutrient balances can vary between years as crop growth fluctuates with rainfall. The natural vegetation is open forest savannah with a trend towards increasing open grassland where rainfall is lowest.

Farming Systems

Northern Nigeria has been classified into three categories of farming system: intensive, less intensive and extensive (Mortimore, 1989). The intensive systems use permanent annual or biannual cultivation with a cropping intensity of more than 60 percent. The low-intensity systems operate a shrub/short-bush - fallow regime and the cropping intensity is 30 - 60 percent. The extensive system is one of long-bush - fallow and uncultivated areas, where the cropping intensity is typically less than 30 percent. The major crops grown are millet, sorghum, groundnut, sesame and cowpea. As the systems are low input, grain yields are about 1 tonne/ha. Ridging ploughs are primarily used for cultivation and hoeing is commonplace. Applications of manure have to be made carefully to avoid "burning" the crops, and additions are 1 - 7 tonnes/ha. Crop residues are collected for fodder or left to be grazed in the fields. Fallowing is practised in the less intensive systems although the land is still "harvested" through grazing and the collecting of wood and other products. Night-parking (or folding) of livestock is the most effective method of providing manure. Legumes such as cowpea are grown, primarily in higher-intensity systems, to provide nitrogen inputs. Inorganic fertilizers are scarce and seldom available at the optimal time.

Study Sites

The study sites cover a range of population density and include the three categories of farming intensity. The soils and crops grown are similar across all categories (Harris, 2000). Initially, CENTURY was parametized with a natural system of grass, woodland and grazing.

A medium-intensity fire was scheduled for every 10 years and a high-intensity fire event occurred every 30 years.

One thousand years ago, slash and burn episodes were commenced and two seasons of millet cultivation incorporated, repeated every 60 years. In the nineteenth century, the frequency of cultivation events was increased to a 30-year cycle, and then once every 15 years by the commencement of the twentieth century.

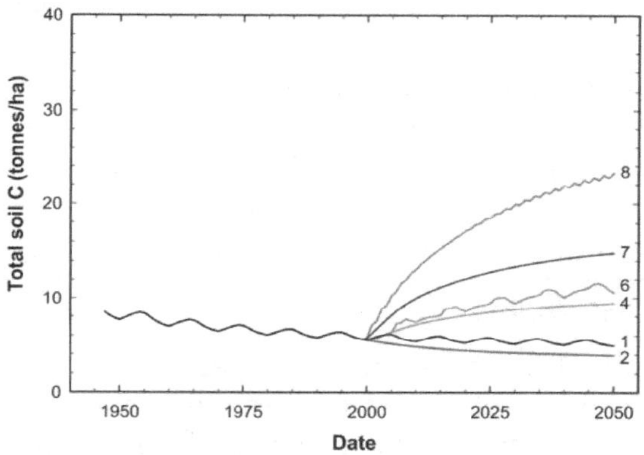

Figure: *Total soil carbon for Futchimiram settlement (CENTURY)*

Futchimiram, Borno State

This is a low-intensity, or extensive, agropastoral system practising shifting cultivation. Some land has now become degraded. CENTURY was run for the last 60 years with alternate five-year cycles of grazing and millet cropping. Crop residues are grazed and there are no other inputs. This current practice produces a gradual and persistent decline in soil C. The estimated level is slightly above the measured value of soil C for cultivated soils that range between 3.5 and 4.4 tonnes/ha. RothC also predicts that the current practice will decrease soil C slightly over the next 50 years

Table: *Total soil carbon for Futchimiram settlement*

Scenario[1]	*CENTURY*			*RothC*		
	2000 (*tonnes/ha*)	*2050*	*% change*	*2000* (*tonnes/ha*)	*2050*	*% change*
1	5.54	4.94	-10.8	5.38	5.18	-3.7
2		3.93	-29.1		3.72	-30.9
5		9.40	69.7		9.70	80.3
6		10.57	90.8		12.45	131.4

Source: CENTURY and RothC.

Table: *Scenarios for modelling land management practices, Futchimiram settlement.*

Scenario	*Land management*
1	Current practice
2	Continuous cultivation
3	Continuous cultivation, no grazing of residues, harvest only grain
4	Inorganic fertilizer only (100 kg/ha urea), no grazing
5	Plant residues average 0.5 tonnes/ha/year, no grazing of residues
6	5-year fallow, 5-year cultivation, 2 applications FYM 3 tonnes/ ha, graze residues
7	Continuous cultivation, FYM 1.5 tonnes/ha/year, graze residues
8	Continuos cultivation, FYM 1.5 tonnes/ha/year Plant residues 0.5 tonnes/ha/year, no grazing

Effect of Loss of Fallow

Removing the fallow period only (Scenario 2) results in a greater decline in soil C over subsequent years, with both models predicting similar reductions. Preventing grazing of the crop residues under these conditions (Scenario 3) has very little effect on soil C.

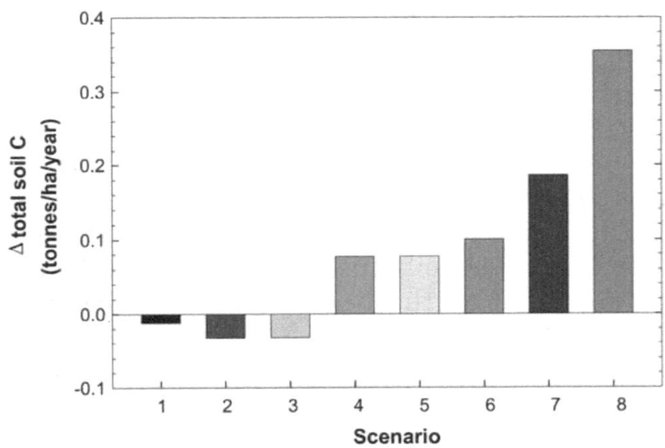

Figure: *Average annual change in total soil carbon for Futchimiram settlement (CENTURY)*

Effect of Organic Inputs

The addition of FYM (two applications of 3 tonnes/ha in each five-year cropping cycle, average 0.6 tonnes/ha/year) has a positive effect on soil C, the two models indicating an increase of 5 - 7 tonnes C/ha over the next 50 years (0.08 tonnes C/ha/year). The outcome of further scenarios using different combinations of fallow and organic inputs. The gradual increase in soil C occurs with FYM, retaining plant residues and no grazing after harvest. In this case, soil stocks rise from 6 to 24 tonnes C/ha over 50 years.

Effect of Inorganic Fertilizer

The use of inorganic fertilizer with no other organic inputs, and no grazing or fallow (Scenario 4), leads to a modest increase in CS (0.08 tonnes C/ha/year).

Summary

This system will not yield a positive soil carbon balance without increased organic inputs. Although inorganic fertilizers can lead to CS, they bring with them a carbon cost that will result in an overall negative balance of the complete carbon budget.

Kaska, Yobe State

This is a low-intensity agropastoral farming system covering lowland, upland and some wetland areas. Only the lowland soils are modelled here. There is intercropping of legumes and grains, and crop residues are fed to livestock. Manure application to fields is low and long bush fallows are used. CENTURY was parametized for the last 50 years with a sevenyear cycle of 4 years of grazing and 3 years of millet - cowpea - millet cropping. FYM (0.75 tonnes/ha) was added in the first year of each cropping cycle. Cultivated soils have carbon contents of 4.5 - 7.0 tonnes/ha. The CENTURY model calculates a current carbon content of 7.7 tonnes/ha, and RothC gives a similar result. Both models suggest that with current practice the system is close to steady state.

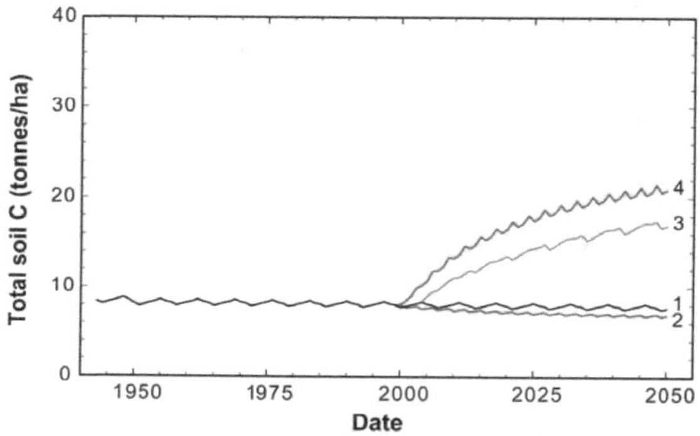

Figure: *Total soil carbon for Kaska settlement (CENTURY)*

Effects of Fallows and Organic Inputs

Removing the fallow from the current practice leads to a slight decline in soil C in subsequent years.

Applying 3 tonnes/ha FYM to each millet crop (average 1.3 tonnes/ha/year over seven-year cycle) will produce a marked increase in soil C representing a CS rate of 0.18 tonnes/ha/year.

This increase will be further enhanced if the fallow is removed, because the manure application rate will now average 2 tonnes/ha/year.

Summary

The scenarios for Kaska illustrate the effect of fallow periods on stocks of soil C. Where the cropping regime is adding very little organic matter to the soil, fallows will often have a positive effect if correctly managed.

However, if the cropping practice is accumulating significant organic matter in the soil, any interruption, such as fallowing, will decrease the overall potential for CS.

Table: *Total soil carbon for Kaska settlement*

Scenario[1]	*CENTURY*			*RothC*		
	2000	*2050*	*% change*	*2000*	*2050*	*% change*
	(tonnes/ha)			*(tonnes/ha)*		
1	7.73	7.64	-1.2	7.33	7.87	7.4
3		16.94	119.1		15.57	112.4

Source: CENTURY and RothC.

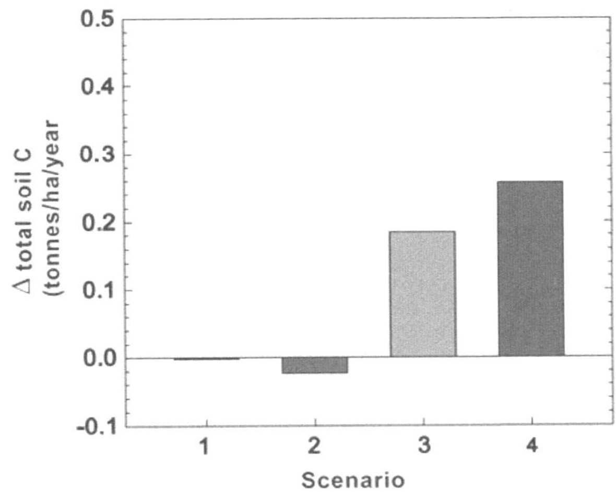

Figure: *Average annual change in total soil carbon for Kaska settlement (CENTURY).*

***Table:** Scenarios for modelling land management practices, Kaska settlement.*

Scenario	*Land management*
1	Current practice
2	Continuous cultivation, millet-cowpea
3	Cultivation-fallow, FYM 3 tonnes/ha, to millet
4	Continuous cultivation, FYM 3 tonnes/ha, to millet

Dagaceri, Jigawa State

This is a region undergoing rapid intensification. It is an agropastoral system with shrub or short-bush fallowing. The length of fallow has decreased as the area of arable land has expanded. Both legumes and grains are grown, but land degradation is a problem. As the fallow period shortens, farmers rely increasingly upon manures to maintain soil fertility. The parametization for CENTURY was initially the same as for Kaska but then, 30 years ago, cropping was increased to five years out of seven. Millet and cowpea are cropped alternately with 1.5 tonnes/ha manure added to each millet crop (average 0.64 tonnes/ha over a seven-year cycle). The additional manure input associated with the increase in cropping intensity results in an increase in soil C, predicted by both models, and this rise continues in subsequent years. Field measurements vary according to previous cropping intensity and range from 3 to 17 tonnes C/ha.

***Table:** Total soil carbon for Dagaceri settlement (CENTURY and RothC)*

Scenario[1]	*CENTURY*			*RothC*		
	2000	*2050*	*% change*	*2000*	*2050*	*% change*
	(tonnes/ha)			*(tonnes/ha)*		
1	12.43	15	20.7	14.69	20.43	39.1
5		22.22	78.8		29.55	101.2

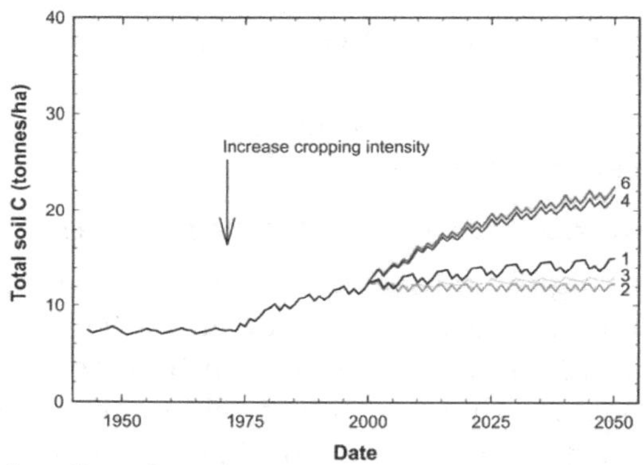

***Figure:** Total soil carbon for Dagaceri settlement (CENTURY)*

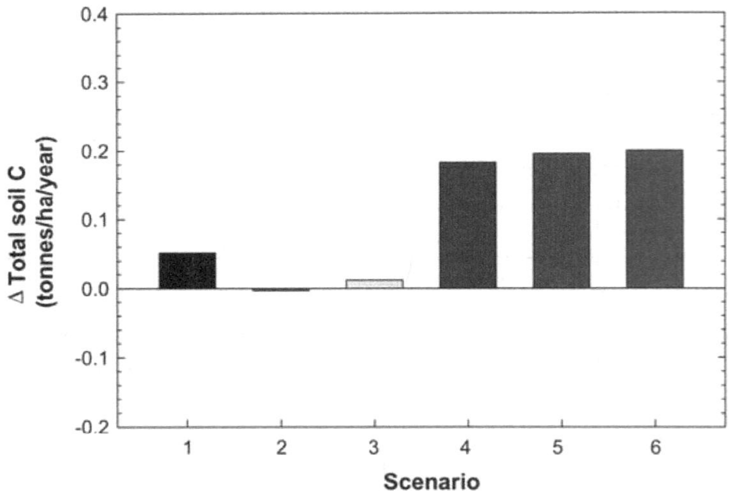

Figure: *Average annual change in total soil carbon for Dagaceri settlement (CENTURY)*

Effects of Fallows and Organic Inputs

The manure input to each millet crop is increased to 3 tonnes/ ha in all scenarios (2 - 6). Carbon sequestration rates of 0.18 - 0.20 tonnes/ha can be achieved, with slight differences depending on whether the fallow is retained and crop residues grazed or ungrazed. However, if fallow elimination is accompanied by harvesting all aboveground material (Scenario 3), CS is virtually halted even if manure application is maintained.

The complete removal of trees from the system results in a net loss of soil C in spite of the increased application of manure.

Summary

This system shows that SOM can be maintained and increased even with intensification of cropping, provided there are legumes in the rotation. However, careful management of crop residues is vital as is the preservation of trees.

iv. Tumbau, Kano Closed - Settlement Zone

This is a highly intensive agricultural area. All the land is cultivated although degradation reportedly affects less than 10 percent of the land area. There is a highly integrated crop and livestock production system with intercropping of legumes, intensive manuring and inorganic fertilizer application. There is virtually no grazing land so animals have to feed on crop residues and fodder from nearby fields.

CENTURY was run for the last 50 years with a millet - cowpea rotation, 6 tonnes/ha of manure applied to the millet (average 3 tonnes/ ha/year) and all aboveground plant material harvested. This system is now close to steady state with soil C stocks at 9.8 tonnes C/ha. This compares with an average of 10.5 ± 1.7 tonnes C/ha measured at cultivated sites. RothC calculates a value of 11.3 tonnes/ha for 2000 but predicts that a higher soil carbon level will be reached by 2050.

Table: *Scenarios for modelling land management practices, Dagaceri settlement.*

Scenario	***Land Management***
1	Current practice
2	Remove trees
3	No grazing of residues, harvest all aboveground
4	Continous cultivation, millet - cowpea
5	FYM average 1.29 tonnes/ha/year,fallow, graze residues, harvest only grain
6	No grazing of residues, harvest only grain

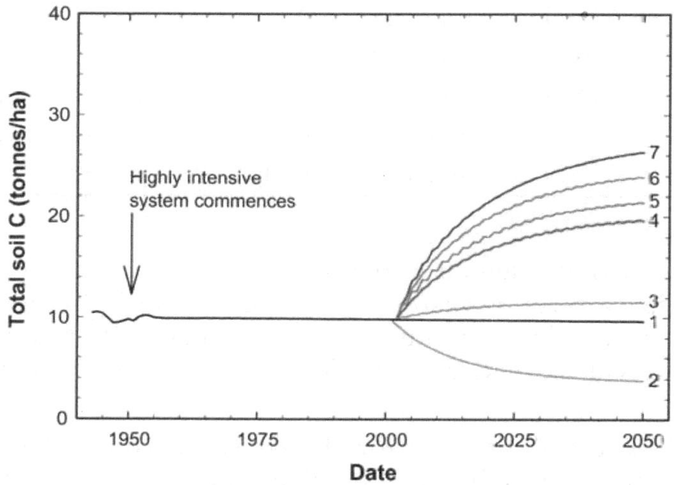

Figure: *Total soil carbon for Tumbau settlement (CENTURY)*

Effects of Organic Inputs

Additional FYM inputs have a marked impact on soil C, especially when they approach the maximum (7 tonnes/ha) normally applied in this region, i.e. an annual input of 6.75 tonnes/ha will result in the sequestration of 0.20 tonnes C/ha/year over the next 50 years. Although adding Nfixing trees and plant residues to the fields will further increase soil C, the requirement of the former for livestock feed may exceed the capacity of the current farming system.

Effect of Inorganic Fertilizer

Replacing the manure input to this system with inorganic fertilizer (urea 100 kg/ha, Scenario 2) results in a large reduction in soil C, with stocks falling by more than 0.1 tonnes C/ha/year.

Table: *Total soil carbon for Tumbau settlement (CENTURY and RothC)*

*Scenario**	*CENTURY*			*RothC*		
	2000	*2050*	*% change*	*2000*	*2050*	*% change*
	(tonnes/ha)			*(tonnes/ha)*		
1	9.82	9.64	-1.8	11.3	13.39	18.5
3		11.54	17.5		14.96	32.4
6		23.9	143.4		31.7	180.5

Table: *Scenarios for modelling land management practices, Tumbau settlement.*

Scenario	*Land management*
1	Current practice
2	Inorganic fertilizer (110 kg/ha urea)
3	FYM 3.75 tonnes/ha/year
4	FYM 6.75 tonnes/ha/year
5	FYM 3.75 tonnes/ha/year, add nitrogen-fixing trees
6	FYM 6.75 tonnes/ha/year, plant residues 2 tonnes/ha
7	FYM 6.75 tonnes/ha/year, harvest only grain

Summary

Provided that adequate organic matter is returned to the soil, these intensive-farming systems should maintain soil C and there is also scope for CS. These results are in agreement with findings in the field that provide no evidence for a decline in SOM in spite of increased cultivation pressure.

However, the ability to realize future CS will depend on a careful balance between cropping and livestock husbandry and the overall capacity of the system. Maintaining crop yields through the application of inorganic fertilizer alone will probably result in substantial losses of SOM.

Conclusions From Northern Nigeria Cases

The modelling of farm data for the drylands of northern Nigeria shows that soil carbon stocks can be increased from a low base with a variety of technologies and practices already available to farmers.

The total amounts of C that can be sequestered with the use of legumes, fallow periods, FYM and retention of plant residues varies between 0.1 and 0.3 tonnes C/ha/ year. This amount rises when trees are also cultivated.

Soil C is lost when only inorganic fertilizers are used to maintain soil fertility - some 0.1 tonnes C/ha/ year in the intense systems of the Kano CSZ. Continuous cultivation results in small year-on-year losses of C where there are no additional inputs of organic matter. In spite of considerable intensification of current systems (shortening of fallow periods), farmers are maintaining carbon stocks. The benefits of maintaining trees in the landscape are shown in these modelled scenarios.

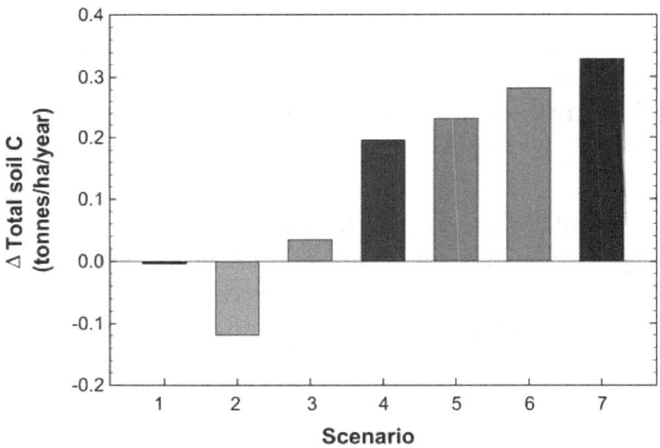

Figure: *Average annual change in total soil carbon for Tumbau settlement (CENTURY)*

Case Study 2 - India - Andhra Pradesh and Karnataka States

More than half of India's farmers of India live in climate regions that can be described as semi-arid. Recent decades have seen increases in crop yields that have been attributed to the green revolution. However, the associated technologies, e.g. irrigation and inorganic fertilizer are expensive, and not readily accessible to the rural poor. These practices may lead to declines in soil fertility, and are also dependent upon fossil-fuel energy (Butterworth, Adolph and Satheesh, 2002). In fact, nearly two-thirds of the arable land in India remains dependent solely upon rainfall for agricultural production. The Natural Resources Institute (the United Kingdom), the Deccan Development Society and the BAIF Institute for Rural Development (India) have studied soil fertility management in Medak District in Andhra Pradesh and in Tumkur District in Karnataka. They have shown that there is an increasing awareness about technologies for maintaining and improving soil fertility, identifying at least 14 different practices from legume cultivation to vermicompost production. Most involve maintaining and increasing the organic matter content of the soil. These studies provide an opportunity to investigate the effect that these soil-fertility improvement techniques may have on CS.

Physical Aspects

Medak District forms part of the tableland of the Deccan Plateau that extends from Andhra Pradesh into Karnataka. The climate consists of a mild winter period (rabi, November - February), a hot and dry summer (March - May) and the southwest monsoon, when more than 80 percent of rainfall occurs (kharif, June - October). Average rainfall is slightly less than 900 mm. The hottest month is May, just before the onset of rains, when the maximum daytime temperature can reach 40 °C. Conversely, night temperatures can drop to 6 °C in December. Moisture availability for crop growth ranges from 120 to 150 d.

The major soil types are Alfisols and Vertisols. The former include red lateritic soils comprising loamy sands, sand loams and sandy clay loams and are usually nonsaline. The Vertisols, black cotton soils, are potentially more productive with a higher waterholding capacity, moderately alkaline and with a highly soluble salt content. They comprise clay loams, clays and silty clays. The organic carbon content of soils in the area is usually 0.5 - 1 percent. The land lies between 500 - 600 m above sea level, and very little natural vegetation remains. The tropical dry deciduous forest has mainly been felled except on protected government land.

Farming Systems

The farming systems in this region have a high degree of integration between livestock, crops and trees (Pound, 2000). There is a huge agrobiodiversity that is combined with off-farm activities, particularly in Medak District. This situation makes for an efficient use of the limited land resources and acts as insurance against unpredictable weather conditions, a frequent problem in drylands. The average farm size is 2.6 ha (Butterworth, Adolph and Satheesh, 2002). The small and marginal farmers can grow at least eight varieties of crops per half hectare. The cattle population has been falling continuously since the 1980s. This has important implications for agriculture, not only because animal wastes are an important source of organic matter for soils, but also because bullocks make a vital contribution as draught animals, not least in the transport of FYM. Tillage is commonly performed with very basic implements. The greatest shift towards mechanized cultivation has been in Karnataka.

The predominant crops grown in Medak District are paddy, sorghum and maize while irrigated sugar cane is an important cash crop. The major crops grown in Karnataka are paddy, sorghum, finger millet, pearl millet, pigeon pea, green gram, and groundnut (Reddy, 2001).

Farmers attach a high priority to maintaining soil fertility while inorganic fertilizers are ranked poorly in terms of maintaining soil quality. However, many farmers use them because organic alternatives are often unavailable and because the inorganic fertilizers are subsidized.

The importance attached to soil fertility can have unforeseen consequences such as the sale of FYM by the poorest farmers to their more affluent counterparts. The ultimate effect of this will be an increase in the fertility and carbon content of some soils while degrading neighbouring areas.

Modelling Soil Carbon in The Study Villages

The CENTURY agro-ecosystem model was parametized using climate and soil data and run to equilibrium commencing with a grass/woodland system to represent the natural vegetation of this region. Low-intensity grazing and fires occurring every 30 years were included in the cycle.

One thousand years ago, the effects of human interference were introduced with slash and burn events together with the cultivation of a grain crop. The frequency of these events was increased slowly and by the start of the twentieth century, cultivation periods of four years out of ten were introduced.

In the mid-twentieth century, one year of grazed fallow was followed by four years of cropping (sorghum - kharif, and cowpea - rabi). Cultivation consisted of hand or bullock ploughing and hand hoeing. Average annual applications of FYM were 2.1 tonnes/ha over the five-year cycle and all aboveground material was harvested. For the final 30 years of the twentieth century, cultivation was adjusted to reflect current practices. The RothC model was run to equilibrium in the mid-twentieth century, and then parametized using current practice and plant residue quantities calculated by CENTURY.

Data on five types of farm and location are presented below:

i. large mixed dryland farm (Lingampally);
ii. small dryland farm, no livestock (Lingampally);
iii. large farm using irrigation (Yedakulapaly);
iv. small mixed drylands farm (Metalkunta);
v. small mixed drylands farm (Malligere);

Analysis of Land Management by a Large Mixed Farm, Lingampally Village, Medak District

This is a holding of slightly more than 5 ha on predominantly Vertisols. Livestock are fed fodder and plant residues from the fields. Consequently, no plant material is returned to the soil. The animals are also grazed on local common land. The livestock provide manure (2 - 3 tonnes/ha/year). In addition, inorganic fertilizer has been used in recent years (30 - 45 kg/ha of di-ammonium phosphate). Crops modelled are sorghum (kharif May - September) and cowpea (rabi October - January). Century predicts that the current farming practice is resulting in a nearly stable soil carbon content of about 19 tonnes/ha declining to 2050 by nearly 1.5 tonnes/ha. RothC shows a lower total soil carbon content of 16 tonnes/ha in 2000, declining to 15 tonnes/ha by 2050.

Effect of Inorganic Fertilizer

If the application of inorganic fertilizer were halted and the quantity of FYM increased to replace the nitrogen input and maintain yields (increase manure application by 0.6 tonnes/ha, Scenario 3), soil C would increase by only 0.85 tonnes/ ha by 2050 and there would be no net gain in soil C.

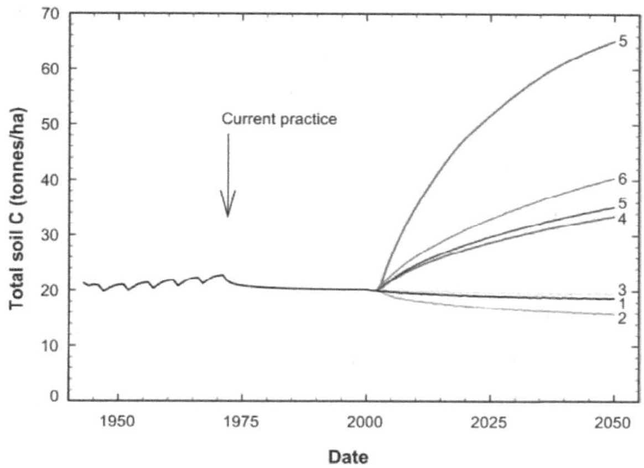

Figure: *Total soil carbon for a large farm (5 ha), Lingampally village (CENTURY).*

Conversely, replacing all organic additions with inorganic fertilizer would result in a decrease in soil C of more than 4 tonnes/ha by 2050, a loss of 0.09 tonnes/ha/ year (Scenario 2). Thus, inorganic fertilizer results in a fall in soil C.

Effect of Farmyard Manure

Doubling the current annual input of FYM to 4 - 6 tonnes/ha without applying inorganic fertilizer has a marked effect on soil C: 0.27 tonnes/ha/year is sequestered in the next 50 years, with the system still not at a steady state. Similarly, RothC shows a continuing rise, reaching 28.3 tonnes/ha by the same date.

Table: *Total soil carbon for a large farm Lingampally village*

Scenario[1]	*CENTURY*			*RothC*		
	2000	*2050*	*% change*	*2000*	*2050*	*% change*
	(tonnes/ha)			*(tonnes/ha)*		
1	20.2	18.8	-7.2	18.3	17.9	-2.3
4		33.5	65.9		28.3	54.2
6		40.4	100		31.0	68.8

Source: CENTURY and RothC.

Effect of Other Organic Inputs

Adding only modest amounts of vermicompost (100 kg/ha) and green manure (250 kg/ha) in addition to the doubled FYM (Scenario 5) has a limited effect, while 2 tonnes/ha plant residues makes a bigger contribution, 0.4 tonnes/ha/year.

Effect of Trees

This farm has a capacity for introducing N-fixing trees such as *Glyricidia,* which creates Scenario 7 when added to the manure and plant residues.

Table: *Scenarios for modelling land management practices, large farm Lingampally village.*

Scenario	*Land management*
1	Current practice
2	FYM 3 tonnes/ha/year
3	FYM 3 tonnes/ha/year, green manure 500 kg/ha/year, vermicompost 250 kg/ha/year
4	As current practice but incorporate crop residues into soil
5	FYM 3 tonnes/ha/year, leave plant residues
6	FYM 3 tonnes/ha/year, plant residues, green manure, vermicompost
7	FYM 6 tonnes/ha/year, plant residues, green manure, vermicompost

After 10 years, the trees are cut annually for wood. The result is a very large increase in soil CS of 0.4 tonnes/ha/year. This increase exceeds that which would be obtained by increasing the manure application by four times.

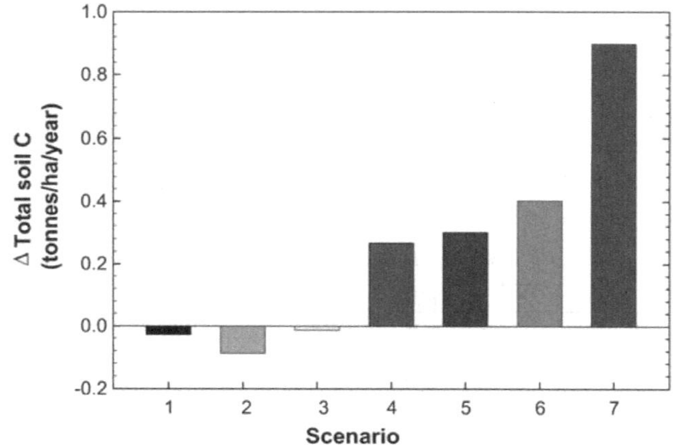

Figure: *Average annual change (over 50 years) in total soil carbon for a large mixed farm, Lingampally village (CENTURY)*

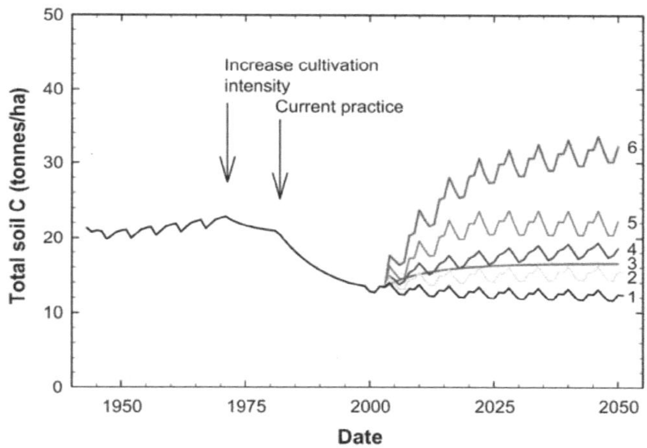

Figure: *Total soil carbon for a small rainfed farm, Lingampally village (CENTURY).*

Summary

The two soil carbon models are in fairly close agreement for this farming system. The current practice is nearly sustainable with only a small decline (< 2 percent) in soil C predicted over the next 50 years. However, cattle are currently being grazed on other land and, consequently, some C is effectively being mined from elsewhere.

A modest increase in organic material would be required to replace the inorganic fertilizers currently used. Further organic inputs could increase soil CS substantially. Introducing trees is likely to have a marked effect on soil C and would simultaneously increase aboveground carbon storage. However, a greater proportion of trees may have been introduced to the model than would be feasible and, therefore, the degree of CS that is practicable for this farming system may be overestimated.

Analysis of Land Management by a Small Rainfed Farm, Lingampally Village, Medak District

This holding is less than 1 ha and the Vertisol has become very degraded. Animals were kept in the past but have been sold and currently sorghum is the only major crop grown. All aboveground material is harvested and removed. Some FYM is applied at different times, equivalent to 3.9 tonnes/ha/year. The CENTURY model shows how current practices have produced a marked decline in soil C, reaching some 13 tonnes/ha in 2000. However the system is reaching steady state and the current land-management practice is not predicted to cause much further decline in soil C in the next 50 years. RothC yields a slightly higher carbon content in 2000 but predicts a slightly greater decline (1.8 tonnes/ha) by 2050.

Effect of Retained Residues

Harvesting only the grain and the returning crop residues to the soil would produce a positive carbon balance, sequestering 0.05 tonnes/ha/year.

Effect of Farmyard Manure

Increasing the input of FYM by 50 percent to 6 tonnes/ha/year would increase soil C by 3.8 tonnes/ ha by 2050. RothC predicts a rise of 3.0 tonnes/ha.

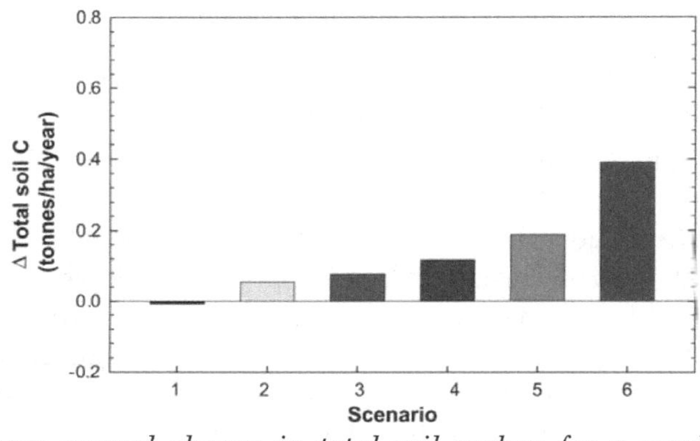

Figure: *Average annual change in total soil carbon for a small rainfed farm, Lingampally village (CENTURY)*

Effect of Legume Crops

Adding cowpea the rotation would sequester 0.12 tonnes/ha/year. Introducing Nfixing trees (cut every 2 years after an initial 10 years) to the sorghum-only system with FYM at 4 tonnes/ha, would yield a sequestration rate of 0.19 tonnes/ha/year. Combining all treatments would lead to a soil CS rate of 0.39 tonnes C/ha/year.

Table: *Total soil carbon for a small rainfed farm, Lingampally village.*

*Scenario**	*CENTURY*			*RothC*		
	2000	*2050*	*% change*	*2000*	*2050*	*% change*
	(tonnes/ha)			*(tonnes/ha)*		
1	12.77	12.41	-2.8	16.07	14.92	-7.2
3		16.58	29.8		19.11	18.9

Source: CENTURY and Roth C.

Table: *Scenarios for modelling land management practices, for a small rainfed farm, Lingampally village*

Scenario	*Land management*
1	Current practice
2	Harvest only grain
3	FYM 6 tonnes/ha/year
4	Add legume (cowpea)
5	Add trees,e.g. *Glyricidia*
6	All additions

Summary

The modelling suggests that the loss of soil C can be reversed for this type of farm, even on degraded soil. In addition to direct organic inputs, this example illustrates the importance of a leguminous crop, and the inclusion of trees such as *Glyricidia.*

Analysis of Land Management for a Large Farm Using Irrigation, Yedakulapaly Village, Medak District

The farm comprises 4.5 ha of irrigated Vertisol. Water is pumped from a borehole. For the purposes of modelling, 5 cm of water is applied when the soil waterholding capacity is < 25 percent during the vegetative growth stage of the crop. Livestock were originally kept but were sold some years ago because there was insufficient labour to collect fodder and tend to the cattle during grazing. Only inorganic fertilizers are now used - 650 kg/ha di-ammonium phosphate and 150 kg/ha of urea annually. Crops are grown in all three seasons (summer - sorghum, kharif - legume, rabi - legume) and all aboveground material is harvested and removed. For the purposes of modelling, cowpea is used for the legume crop.

Century calculates that current practices have depleted soil C to 13.26 tonnes/ha and predicts no further reduction in the next 50 years.

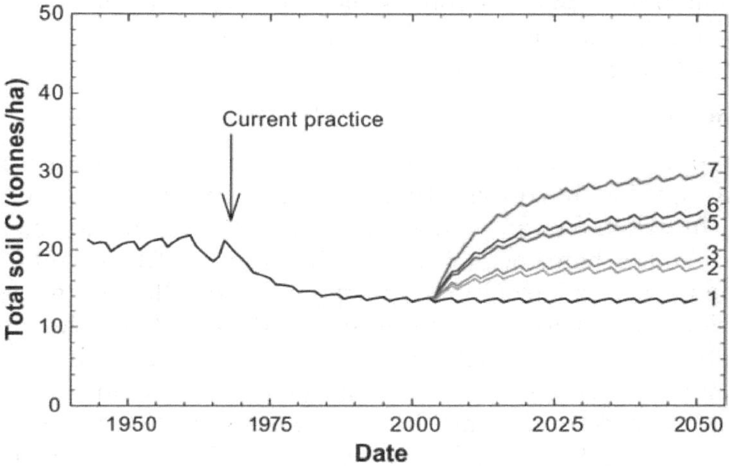

Figure: *Total soil carbon for a large farm using irrigation and cultivating three crops per year, Lingampally village (CENTURY)*

Effect of Organic Inputs

Adding 3 tonnes FYM/ha/year as per Scenario 2 would increase soil C by 4.4 tonnes/ha to 2050, sequestering 0.09 tonnes/ha/year. Including 500 kg/ha/year of green manure and 250 kg/ha/year of vermicompost would increase soil organic C by a further 1 tonne, sequestering 0.11 tonnes/ha/ year.

Table: *Scenarios for modelling land management practices, large farm using irrigation Lingampally village*

Scenario	*Land management*
1	Current practice
2	FYM 3 tonnes/ha/year
3	FYM 3 tonnes/ha/year green manure 500 kg/ha/year, vermicompost 250 kg/ha/year
4	As current practice but incorporate crop residues into soil
5	FYM 3 tonnes/ha/year, leave plant residues
6	FYM 3 tonnes/ha/year, plant residues, green manure, vermicompost
7	FYM 6 tonnes/ha/year, plant residues, green manure, vermicompost

A similar increase in soil C can be obtained with no extra organic additions if only the grain is harvested and the crop residues are returned to the soil. When 3 tonnes FYM/ha/year is applied and only the grain harvested, soil C increases by more than 10 tonnes/ha by 2050, equivalent to 0.20 tonnes/ha/year. Applying all the organic inputs and increasing the FYM component to 6 tonnes/ha would sequester 0.32 tonnes/ha/year.

Summary

A combination of leaving crop residues and adding manure could make a significant contribution to soil C in this kind of farming system, in which current practices have already caused a significant decline in soil carbon stocks. However, inorganic fertilizer and irrigation continue to have a carbon cost, resulting in smaller increases under the various scenarios. Additions of FYM can offset the nutrients applied in inorganic fertilizers. For example, removing inorganic fertilizer from Scenario 6 maintains yield without affecting soil C. Removing irrigation from the above scenario does not have a detrimental effect on soil C, but nor is there an apparent effect on yield.

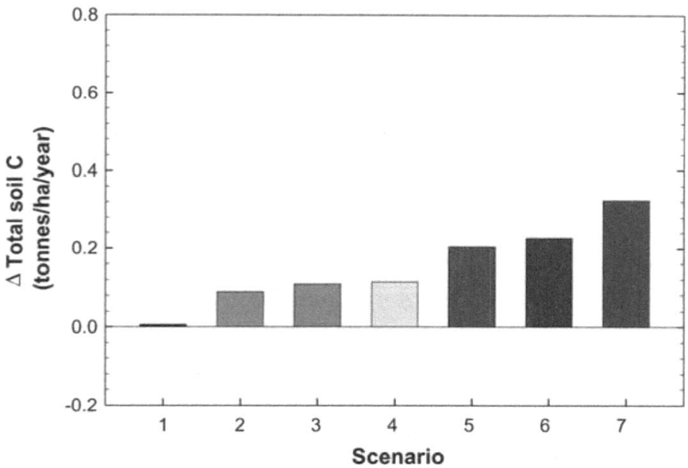

Figure: *Average annual change in total soil carbon for a large farm using irrigation, Lingampally village (CENTURY).*

Analysis of Land Management for a Small Mixed Crop and Livestock Farm, Metalkunta Village, Medak District

This landholding of 2.7 ha has a very large variety of crops on lateritic red soil. Livestock are kept, a good cultivation pattern is practised and there are also many trees present around the farm. All aboveground produce is harvested so that crop residues can be fed to the animals. Cattle are also grazed on nearby land and fodder is brought in. The average yearly manure application is 2 tonnes/ha, and vermicompost is added to the land.

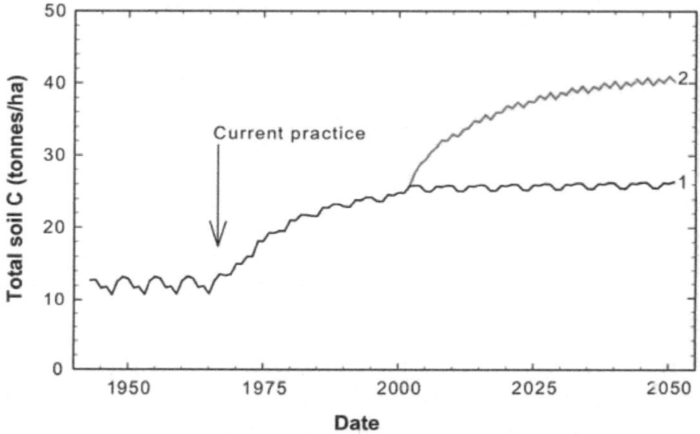

Figure: *Total soil carbon for a small mixed crop and livestock farm, Metalkunta village (CENTURY)*

Century was parametized with a sorghum - cowpea - millet - cowpea - maize - cowpea rotation to reflect the varied cropping pattern. The model was run with this scenario from the late 1960s. The model suggests that this system of land management would have approximately doubled the soil carbon content to nearly 25 tonnes/ha by 2000 and would reach 26 tonnes C/ha by 2050. RothC predicts 25 tonnes C/ha in 2000 rising to 29.1 tonnes/ha in 2050. This mixed cropping and livestock system has a substantial positive impact on soil C.

Table: *Total soil carbon for a small mixed crop and livestock farm, Metalkunta village*

*Scenario**	*CENTURY*			*RothC*		
	2000	*2050*	*% change*	*2000*	*2050*	*% change*
	(tonnes/ha)			*(tonnes/ha)*		
1	24.9	26.1	5.2	25.0	29.2	16.7
2		40.9	64.7		49.2	96.6

Source: CENTURY and RothC.

Table: *Scenarios for modelling land management practices, small mixed crop and livestock farm, Metalkunta village*

Scenario	*Land management*
1	Current practice
2	Leave plant residues

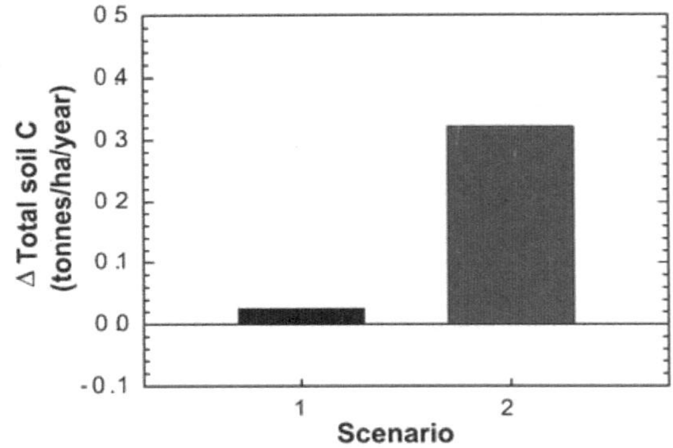

Figure: *Average annual change in total soil carbon for a small mixed crop and livestock farm, Metalkunta village (CENTURY).*

Effects of Retaining Crop Residues

Returning crop residues to the soil would greatly increase the carbon content, sequestering 0.3 tonnes C ha/year in the next 50 years, with the soil carbon content reaching 40.9 tonnes/ ha by 2050. RothC calculates that leaving crop residues would increase soil C to 49.2 tonnes/ha.

Summary

The farmer is currently practising very good land management, and the models calculate that the soil carbon content has already been substantially increased. However, this is being accomplished in part by grazing cattle on other lands and bringing in fodder. This means that C is effectively being mined from elsewhere. Thus, although returning crop residues to the soil makes a further significant contribution to soil C, this would require importing more fodder to substitute for the plant residues, which again implies mining C.

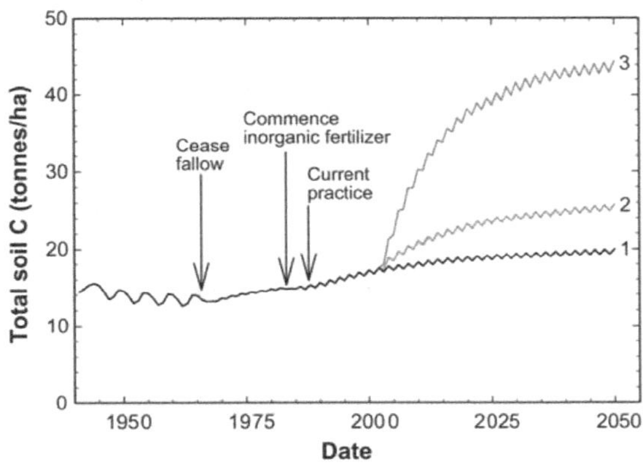

Figure: *Total soil carbon for a small farm, Malligere village, Tumkur District (CENTURY)*

Table: *Total soil carbon for a small farm, Malligere village, Tumkur District.*

*Scenario**	*CENTURY*			*RothC*		
	2000	*2050*	*% change*	*2000*	*2050*	*% change*
	(tonnes/ha)			*(tonnes/ha)*		
1	17.24	19.97	15.8	19.82	24.32	22.7
2		25.73	49.2		34.2	72.6

Source: CENTURY and RothC.

Table: *Scenarios for modelling land management practices, small farm, Malligere village*

Scenario	*Land management*
1	Current practice
2	Replace inorganic fertilizer with FYM
3	Add trees, *Glyricidia*

Analysis of Land Management for a Small Farm, Malligere Village, Tumkur District Karnataka State

This farm covers 2 ha on a red sandy soil. Tractors are hired for transport, livestock are kept and allowed to graze the crop residues, but additional fodder is brought in. The kharif crops modelled are sorghum and millet and the rabi crop is cowpea. On average, 3 tonnes/ ha/ year FYM is applied. Inorganic fertilizers are also used: 75 kg/ ha/ year of di-ammonium phosphate and 75 kg/ha/year of urea.

Modelling with CENTURY shows that current practices are increasing soil C and that this will continue to rise by a further 2.7 tonnes/ha in the next 50 years, reaching almost 20 tonnes C/ha in 2050. RothC also shows that the current practice is increasing soil C but shows an enhanced effect. Soil C in 2000 is predicted to be 19.8 tonnes/ha, rising to 24.3 tonnes/ha in 2050.

Effect of Organic Additions

Inorganic fertilizer has a carbon cost, so replacing the N supplied by the inorganic fertilizer with manure (annual addition 6.3 tonnes/ ha) enhances the CS rate to 0.17 tonnes/ ha/year. The soil carbon level reaches 25.7 tonnes/ha by 2050. RothC predicts a much bigger effect on SOC of replacing the fertilizer with FYM: soil C is predicted to rise to 34.2 tonnes/ha by 2050.

Effect of Trees

Adding trees such as *Glyricidia* to the system (Scenario 3), which are cropped annually for wood after ten years, makes a very large difference, increasing CS to 0.54 tonnes/ ha/year.

Summary

Both models show similar trends, with a switch from inorganic to organic fertilizer increasing soil C significantly. RothC predicts a larger increase. Cessation of inorganic fertilizer use would remove the negative carbon balance associated with its production, and the addition of trees may be sufficient to offset the inputs of C that arise from fodder grown outside the farm system. However, if tractors are used to transport fodder, the carbon budget for the system will probably be negative.

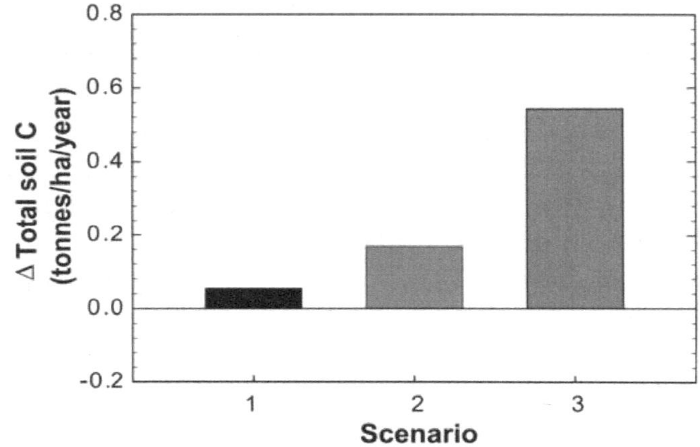

Figure: *Average annual change in total soil carbon for a small farm, Malligere village, Tumkur District (CENTURY)*

Conclusions From India Cases

The modelling of farm data for the drylands of India shows that soil carbon stocks can be increased with a variety of technologies and practices available to farmers.

It also shows that some practices result in substantial declines in carbon stocks, particularly the use of inorganic fertilizer as the sole source of nutrients, and the continuous cultivation of cereals. In the large (5 ha) mixed farm, inorganic fertilizer results in the loss of 0.1 tonnes/ha/year, whereas the use of either FYM, green manures, vermicompost and/or plant residues produces increases of 0.2 - 0.4 tonnes/ha/year. The use of agroforestry substantially increases belowground CS to 0.9 tonnes/ha/year.

The models show substantial declines in soil C on the small farms that cultivate only sorghum and those that intensively cultivate three irrigated crops per year - a loss of 5 tonnes C/ha in a 25-year period. However, it appears that these falls can be reversed within 5 - 10 years with the adoption of legumes in rotations, the addition of FYM manure and cultivation of trees.

The small mixed cropping farm, with cereals and cowpea rotations and livestock, increases soil C from a stock of about 13 tonnes C/ha to 24.8 tonnes C/ha in 25 years. It can increase this to more than 40 tonnes C/ha in the next 50 years if plant residues are added to the soil.

For full accounting of the C used or sequestered in these farms, it is important to consider the high energy cost of nitrogen fertilizer manufacture (65.3 MJ/kg for N, 7.2 MJ/kg for phosphorus, 6.4 MJ/kg for potassium) (Pretty *et al.*, 2002), the use of mechanized operations on farms, the cost of irrigation and mechanical transport, and the issue of the transfer of C in feed or livestock themselves from one farm to another or from grazing areas to cropped fields.

There are clear benefits for farmers and soil C if leguminous crops are included in rotations and in agroforestry systems.

Case Study 3 - Kenya - Makueni District

Arid and semi-arid lands occupy about two-thirds of Kenya (FAO, 1999b). The major factors limiting crop growth and production in these areas of Kenya are erratic rainfall, poor husbandry and declining soil fertility caused by continuous cultivation (Kenya Agricultural Research Institute, 1999). Drylands Research has examined the sustainability of farming systems in semi-arid Kenya, recording physical aspects of the environment together with land-management practices in four villages of Makueni District (Mbuvi, 2000). These span a range of settlement times from the 1950s to the 1970s. A major problem for farming in this region has been the frequency with which it has been affected by drought. Droughts reduce the returns that farmers receive and, consequently, provide little incentive for investment in soil fertility.

Physical Attributes

The climate of this region is characterized by two rainy periods: the short rains, which deliver most precipitation in October - December; and the long rains of March - May. However, superimposed on the average annual rainfall (600 - 670 mm) is the periodic occurrence of drought (Gichuki, 2000). Annual mean temperature is in the range of 21 - 24 °C. Elevation is 800 - 1 600 m, and the natural vegetation is grassland and dense shrub-land or woodland. Fires have affected the area in the past and the grassland is used for grazing. The soils are mostly Ferralsols (Rhodic and Xanthic) and are naturally low in phosphorus (Mbuvi, 2000).

Farming Systems

Annual or multiple cropping is practised with occasional incorporation of one-year fallows, although the latter is becoming less common. Each year covers two cropping seasons. The main crops are maize and pulses, with millet and sorghum recommended as drought crops. Yields vary considerably between years depending on rainfall; on average maize will yield 1 tonne/ha, although some modern varieties yield 4 tonnes/ha of grain (Mbogoh, 2000).

The animal population is not large because it is difficult to supply adequate feed during periods of drought. Consequently, manure is in short supply and highly valued. Scarcity means that its application is often rotated. Little investment is made in the grazing lands, and the animals are usually kept in pens during the wet season. Very little fertilizer is used, especially as drought causes "burning" of the crops. Crop residues are burnt, fed to animals or ploughed into the soil. Tillage is accomplished by using simple ridging ploughs pulled by oxen; hand hoeing and digging are also common. Labour is in short supply and a constraint for farmers. A range of crop-residue management and tillage techniques are used to conserve moisture and to protect against soil erosion (Pretty, Thompson and Kiara, 1995; Gichuki, 2000).

Woodland is being cleared, although selective clearance is often practised to save those species that provide useful products. Most farmers also plant trees, especially fruit trees and others such as mulberry for silk production (Gichuki, 2000). The study villages lie along a gradient of precipitation, decreasing from Kymausoi through Kaiani and Darjani to Athi Kamunyuni. CENTURY was run to equilibrium using a grassland - tree scenario with grass fires every ten years and major fires every 30 years.

Darjani

The primary crops for this settlement are millet, sisal, cowpea and sorghum and the natural pasture can support low intensity grazing. Settlement took place in the 1960s. The average soil carbon content of bush soils is 37.6 ± 7.5 tonnes/ha, and for cultivated soils it is 33.7 ± 2.8 tonnes/ha (Mbuvi, 2000).

Once run to equilibrium, CENTURY gave a value for soil C of 31.0 tonnes/ha in 1964. This is just within the range expected for bush soils. The model was then run for a further 35 years, reproducing the current land management. This commenced with a six-year cycle of alternate maize and millet crops for four years followed by fallow for two years. Cultivation was accomplished using a steel ridging plough and hand weeding was included. Crop residues were burnt. In 1971, FYM additions commenced, averaging 1.5 tonnes/ha over the six-year cycle as soil fertility and conservation measures began to be adopted. In 1983, the fallow period was reduced to one year out of six, and by 2000 the modelled soil C was 24 tonnes/ha. This reduction of 7 tonnes/ ha from the level modelled for the uncultivated bush soil is greater than the decline of 5 tonnes/ha measured in the field. Continuation of this landmanagement practice is predicted to lead to a further loss of 1.3 tonnes C/ha by 2050.

RothC was run to equilibrium using the current level of bush soil C and then used to predict the effect of land management using plant inputs calculated by CENTURY. RothC predicted soil C to be 34.6 tonnes/ha in 1999, which is within 1 tonne/ha of the measured value. In the next 50 years, RothC predicts a further decline of 0.6 tonnes C/ha.

Effect of Organic Inputs

Adding an average 4.5 tonnes/ha/ year FYM over the 6-year croppingfallow cycle leads to a CS rate of 0.5 tonnes/ha/year. If the crop residues are returned to the soil rather than burnt, a further 4.9 tonnes C/ha can be accumulated by 2050, representing a sequestration rate of 0.6 tonnes/ha/year. RothC predicts similar increases. Increasing the FYM input to 6.75 tonnes/ha/year will increase the sequestration rate to 0.9 tonnes/ ha/year.

Effect of Inorganic Fertilizer

Replacing all organic inputs with inorganic fertilizer (100 kg N/ ha/ year, Scenario 4) results in only a moderate increase in soil CS (0.09 tonnes/ha/year). However, the quantity of N applied is equivalent to about five times that which is added in FYM for the current practice scenario. If the fallow period is removed and only inorganic fertilizer added (Scenario 3), the system behaves very much like the current practice scenario in spite of the additional N.

Summary

The importance of FYM as an organic input is demonstrated, as is the return of crop residues to the soil rather than burning them. Although addition of inorganic fertilizer can increase CS, the increase in C per unit of N added is much less efficient than if FYM is added. The indirect energy embodied in manufactured N fertilizer is an additional carbon cost.

Kaiani

The settlement of Kaiani has a farming system similar to that at Darajani. CENTURY underestimated the current bush content of soil C, yielding nearly 32 tonnes/ha compared with measured values of 41.5 ± 3.1 tonnes/ha.

CENTURY was then run with a scenario to reflect the last 40 years of cultivation using a millet - cowpea system with grazing of plant residues and 4.5 tonnes/ha manure applied once in the six-year cycle. There was one year of fallow.

Cultivation is calculated to have reduced soil C to 28.1 tonnes/ha by 2000, which compares with a measured value for cropped soil of 30.5 ± 4.8 tonnes/ha. Soil C is then predicted to decrease by slightly more than 1 tonne/ha in the next 50 years.

Table: *Total soil carbon for Kaiani settlement*

*Scenario**	*CENTURY*			*RothC*		
	2000	*2050*	*% change*	*2000*	*2050*	*% change*
	(tonnes/ha)			*(tonnes/ha)*		
1	28.07	26.81	-4.5	42.32	43.89	3.9
2		41.4	47.5		62.82	48.8
3		42.73	52.2		65.16	54.3

Source: CENTURY and RothC.

Table: *Scenarios for modelling land management practices, Kaiani settlement.*

Scenario	*Land management*
1	Current practice
2	FYM 2 tonnes/ha/year
3	FYM 2 tonnes/ha/year, plant residues 0.3 tonnes/ha/year

Table: *Total soil carbon for Kymausoi settlement*

*Scenario**	*CENTURY*			*RothC*		
	2000	*2050*	*% change*	*2000*	*2050*	*% change*
	(tonnes/ha)			*(tonnes/ha)*		
1	17.76	16.82	-5.3	35.69	34.79	-2.5
2		20.18	13.6		43.4	21.6
3		27.7	56.0		58.86	64.9

Source: CENTURY and RothC.

After parametization with the carbon content for bush soils, RothC shows very little effect of cultivation when using plant inputs calculated by CENTURY. Unlike CENTURY, this model predicts a slight rise in soil C by 2050.

Effect of Organic Inputs

Modest increases in the application rate of FYM (0.75 tonnes/ha/ year to 2 tonnes/ha/year) result in quite marked accumulation of soil C. Adding additional plant residues further increase the sequestration rate to 0.3 tonnes/ha/year. RothC predicts very similar proportional increases in soil C.

Summary

The models suggest that the current system is at or near steady state. It is possible to achieve reasonable rates of CS (0.3 tonnes/ha/ year) with modest increases in organic manure inputs.

Kymausoi

Kymausoi is situated in a marginal cotton zone where maize, pigeon pea and sisal are also grown. Cattle ranching is practised and the village was settled in the 1950s. The average soil carbon content for bush soils here is 38.4 ± 4.8 tonnes/ha, while for cultivated soils the average is 33.5 tonnes/ha with a range of 17.4 - 38.9 tonnes/ha. The CENTURY agro-ecosystem model was run to equilibrium and gave a bush soilcarbon level of 28.5 tonnes/ha in 1955, which is below the average concentration for soils in this area. The model was then run to reflect the farming system since settlement commenced. This includes continuous maize cropping with grazing of the crop residues, and an average manure application rate of 0.75 tonnes/ha/ year. Soil conservation and fertility management commenced in the late 1960s, the average annual manure application rate being increased to 1 tonne/ha to reflect this. CENTURY estimated soil C to be 17.8 tonnes/ ha by 2000, which is at the lower end of measured values. A decline of less than 1 tonne C/ha is predicted to occur in the next 50 years.

Table: *Scenarios for modelling land management practices, Kymausoi settlement.*

Scenario	*Land management*
1	Current practice
2	Plant residues 0.3 tonnes/ha/year
3	FYM 1.5 tonnes/ha/year, plant residues 0.6 tonnes/ha/year
4	FYM 1.5 tonnes/ha/year, plant residues 0.6 tonnes/ha/year, legume (cowpea)
5	FYM 2 tonnes/ha/year, plant residues 0.6 tonnes/ha/year, legume (cowpea)
6	FYM 4 tonnes/ha/year, plant residues 0.6 tonnes/ha/year, legume (cowpea)

***Table:** Total soil carbon for Athi Kamunyuni settlement*

*Scenario**	*CENTURY*			*RothC*		
	2000	*2050*	*% change*	*2000*	*2050*	*% change*
	(tonnes/ha)			*(tonnes/ha)*		
1	23.75	23.45	-1.3	29.68	28.29	-4.7
2		29.09	22.5		31.40	5.8
3		29.77	25.3		42.76	44.1

Source: CENTURY and RothC.

After being run to equilibrium for current bush soils and then parametized with plant inputs from CENTURY, RothC calculated a reduction of 3.4 tonnes/ha in soil C resulting from cultivation. Like CENTURY, it suggests a further decline of less than 1 tonnes/ha by 2050.

Effect of Organic Inputs

Adding additional plant residues and increasing the FYM application rate to 1.5 tonnes/ha/year will increase soil carbon levels by at least 50 percent.

Effect of Legume Crop

Introducing a leguminous crop into the system, such as cowpea, can make a significant improvement in CS, increasing the rate from 0.2 to 0.3 tonnes/ha/year at the same rates of FYM application.

Summary

This example illustrates the advantage of including leguminous crops in the rotation. A farm system using 4 tonnes/ha/year of FYM, maintaining plant residues in the field, and with a legume in the rotation, can accumulate 0.7 tonnes/ ha/year. The current practice is reducing soil carbon stocks.

Athi Kamunyuni

This village is situated in a lowland agro-ecological zone. Ranching will only support a very low grazing density and the area is at the limit for rainfed production of millet, cowpea and sisal. The village was established in the 1970s. CENTURY was parametized for the commencement of cropping/grazing during the last 30 years. Millet was grown and an average of 0.75 tonnes/ha manure applied during the last six years.

The value for soil C modelled by CENTURY matched the current bush level (33.2 ± 3.2 tonnes C/ha). When settlement commenced, CENTURY calculated a decline in soil C of some 9 tonnes/ha over 25 years. Again, this is close to the actual measurements that show cultivated soils on average contain 24.0 ± 3.0 tonnes C/ha. Continuing this scenario into the twenty-first century suggests that the system has almost reached a new steady state. RothC calculates a smaller effect of cultivation on soil C, predicting a value of 29.7 tonnes C/ ha in 2000, and a slight decline over the next 50 years.

Table: *Scenarios for modelling land management practices, Athi Kamunyuni settlement.*

Scenario	*Land management*
1	Current practice
2	Grazing
3	FYM 1.25 tonnes/ha/year
4	FYM 2.25 tonnes/ha/year
5	FYM 2.27 tonnes/ha/year, fallow reduced to 1 year
6	FYM 3.3 tonnes/ha/year
7	FYM 3.9 tonnes/ha/year, plant residues 0.3 tonnes/ha

Effect of Organic Inputs

Increasing the current average application rate of FYM from 0.75 tonnes/ha/year to 1.25 or 2.25 tonnes/ha/year (Scenarios 3 and 4) increases rates of CS by 0.12 - 0.37 tonnes/ha/year. Further additions of organic inputs, if available, could yield an increase in CS rates of up to 0.6 tonnes/ha/year.

Effect of Fallows

Reducing the fallow period from two years to one year (Scenario 5) has very little effect on soil C providing the organic inputs are maintained. Using this location solely for ranching would result in an increase in soil C as the system returned to similar conditions that existed before cultivation commenced (Scenario 2). The rate of CS is very similar to Scenario 3, where an average of 1.25 tonnes/ha FYM was added annually. RothC predicts a much smaller increase in soil C on return to ranching. This reflects the fact that this model initially calculated a much smaller decline in total soil C following the commencement of cultivation.

Summary

The level of soil C in this recently settled system could be restored to pre-settlement levels and then raised higher by the addition of FYM. Conditions at this location are not ideal for cropping and a return to a grazing-only system should also restore soil C to its pre-settlement level.

Conclusions from Kenya Cases

The modelling of farm data from four communities in the semi-arid Makueni District again shows that carbon stocks can increase when a variety of technologies and practices already available to farmers are used.

Modest inputs of organic material in the form of FYM and plant residues can lead to CS, particularly where systems are currently at or near steady state for soil carbon stocks.

Removal of fallow periods from existing systems results in losses of 0.1 tonnes C/ha/year. Inorganic fertilizers are again an inefficient choice for plant nutrients when soil C is a concern. Burning plant residues is not desirable.

The combination of legumes in rotations, 2 - 4 tonnes/ha/year of FYM in addition to 0.6 tonnes/ha/year of plant residues, results in the highest rate of CS in all the dryland cases - 0.7 tonnes C/ha/year. At lower levels of FYM but with maintenance of fallows, combined with legumes in rotation, increases in soil C of 0.3 - to 0.4 tonnes/ha/year can be achieved.

Case Study 4 - Argentina - Tucuman, Catamarca and Cordoba Provinces

In recent years, Argentina has experienced a rapid growth in the adoption of reduced and no-tillage systems, especially in dryland regions. This change has been brought about by a deterioration in soil quality and associated crop yields. Many local soils are not suited to the heavy tillage and cropping practices introduced by European settlers.

The Argentine Pampa now has very little natural vegetation. Xerophitic vegetation such as *Prosopis algarrobilla* and *Larrea divaricata* can still be found in the most arid areas. Agricultural practices commenced with the arrival of colonists in the sixteenth century. Ungulates were introduced to graze the grasslands, which have now been mostly re-sown. Very few trees remain except around farmsteads. Wheat was initially cultivated and row-crop production has increased with time. In many parts, grazed pasture was dominant until the 1990s but since then there has been a marked increase in the cultivation of summer annuals, such as maize, sunflower and soybean (Diaz-Zorita, Duarte and Grove, 2002). The Argentine Pampa has been recognized as a region with potential for increased production, if soils can be improved (Alvarez, 2001).

Crop yields have declined in many areas. These declines have been correlated closely with a reduction in SOM content (Diaz-Zorita, Duarte and Grove, 2002). This has prompted the need for change in existing land-management practices. The negative effects of heavy tillage on SOM led to the commencement of no-tillage experiments in the 1960s, in an attempt to produce a more sustainable agricultural system. Now some 13 million ha, or about half of the agricultural area in Argentina, is under some form of reduced-tillage system. Fertilization of crops is primarily achieved through the use of inorganic fertilizers, with organic material tending to be conserved for use in horticultural farming systems.

No-tillage Modelling Studies

Three case studies are reviewed to model soil C under a variety of conventional and no-tillage systems in Tucuman, Catamarca and Cordoba Provinces.

Monte Redondo, Tucuman Province

This is a semi-arid area that naturally supports xerophitic vegetation. The agricultural practices include grazed prairie lands and row cropping, the two systems often being rotated. The studied site consists of cropping for seven years followed by four years of prairie grassland. The crop sequence is wheat/soybean, maize, soybean, wheat/ soybean, maize, soybean, wheat, and four years of prairie. Both conventional tillage and no-tillage cultivation are practised. In the tillage system, disc and chisel plough are used for soil preparation, while the no-tillage system uses the same cropping sequence without tillage.

After an equilibrium phase of grassland and trees, with fire every 60 years, CENTURY was parametized to run with improved prairie grassland from the midnineteenth century. The model predicts that the prairie system is losing C at the rate of 0.06 tonnes/ha/year under the current regime, but at stocks of 55.4 tonnes C/ha, it is overestimating the current measured level of 48.8 tonnes C/ha.

Cultivation is scheduled to commence in the 1950s. Fertilizer applications of 110 kg/ha urea begin in 1980. Two scenarios compare the effect of conventional tillage with that of no-tillage. The model predicts a consistent difference between the two systems of less than 3 tonnes C/ha. This is less than the difference of 6 tonnes C/ha currently measured in the field. CENTURY predicts that this difference between the two systems will be maintained into the future, although both systems continue to lose soil C. The pattern of fall and rise in the soil C curve occurs because the crop part of the rotation results in soil C loss, while the return to prairie increases soil C.

Table: *Total soil carbon for Monte Redondo*

Scenario*	CENTURY			RothC		
	2000	2050	% change	2000	2050	% change
	(tonnes/ha)			(tonnes/ha)		
Prairie	55.40	54.50	-1.6	43.95	43.21	-1.7
Tillage	40.43	39.27	-2.9	32.57	27.58	-15.3
No-tillage	43.64	42.23	-3.2	35.33	33.58	-5.0
5		48.23	4.4		41.89	18.6

Source: CENTURY and RothC.

Table: *Scenarios for modelling land management practices, Monte Redondo.*

Scenario	Land management
1	Conventional tillage, inorganic fertilizer
2	No-till, inorganic fertilizer
3	No-till, FYM 1.5 tonnes/ha/year, inorganic fertilizer
4	No-till, green manure 10 tonnes/ha/crop, inorganic fertilizer
5	No-till, FYM 1.5 tonnes/ha/year, green manure 10 tonnes/ha/crop, no inorganic fertilizer
6	No-till, FYM 3.3 tonnes/ha/crop, no inorganic fertilizer

RothC calculates lower levels of soil C but predicts a greater differential of 6 tonnes/ha by 2050, mainly through a higher loss of C from the tilled system.

Effect of No-tillage and Organic Additions

Additions of green manure (10 tonnes/ha/crop) or FYM (1.5 tonnes/ha each cropping year) to the no-tillage system both lead to increases in CS. A combination of these inputs without inorganic fertilizer yields a similar result.

Cessation of inorganic fertilizer usage and using FYM as a replacement source of N results in the highest rate of CS (Scenario 6), 0.1 tonnes/ha/year.

Summary

Both models register the improvement that no-tillage has on soil carbon content. However, if the decline in soil carbon content is to be reversed, additional inputs of organic matter are required - either from FYM or use of green manures in the rotation. An increase in prairie in the rotation will also increase soil carbon stocks.

Santa María River Valley, Catamarca Province

This is an arid region, with mean annual rainfall of 400 mm and temperatures in the ranging of 7 - 32 °C. The native vegetation is xerophitic consisting of creosote bush scrub (*Larrea divaricata*) and trees such as *Prosopis algarrobilla*. In cultivated areas, vines are grown and crops include alternate plantings of red pepper and barley. Measured soil carbon stocks are high in this district at 3.9 percent.

Table: *Total soil carbon for Santa Maria*

*Scenario**	*CENTURY*			*RothC*		
	2000	*2050*	*% change*	*2000*	*2050*	*% change*
	(tonnes/ha)			*(tonnes/ha)*		
Prairie	37.11	35.75	-3.7	70.68	74.71	5.7
Tillage	21.39	18.11	-15.3	51.46	47.36	-8.0
No-tillage		24.00	12.2		53.53	4.0
3		30.49	42.5		58.71	14.1

Source: CENTURY and RothC.

After reaching equilibrium with natural vegetation, CENTURY was run with a prairie scenario from the mid-nineteenth century and cultivation commenced in the late 1950s using conventional tillage (disc plough).

A cotton - barley rotation commenced in 1980. CENTURY calculates that soil C had decreased to two-thirds of its initial value by 2000. Measurements in the region confirm declines in soil C in cultivated areas of 33 - 66 percent. Therefore, the model estimate is at the top of this range. However, the system is predicted to reach a new steady state and the decline in soil C is estimated to be 3 tonnes/ ha over the next 50 years.

RothC, parametized with the higher, measured level of soil C and using quantities of plant inputs calculated from CENTURY shows a proportionately smaller effect of the current cultivation practice on soil C.

Effect of no-tillage

Adopting a no-tillage system not only halts the loss of soil C predicted by CENTURY. It also leads to a low sequestration rate of 0.05 tonnes/ha/ year in the next 50 years. The RothC model is less sensitive to this scenario.

Effect of Organic Additions

Additions of FYM and green manure both lead to marked increases in CS rates of 0.18 - 0.25 tonnes/ha/year. RothC again shows a smaller effect.

Effect of Inorganic Fertilizer

Replacing inorganic fertilizer with FYM promotes CS (0.22 tonnes/ha/year). Combining the green manure and FYM applications gives the best carbon accrual rate, 0.29 tonnes/ha/year. Inorganic N fertilizer is inefficient owing to the high energy cost of manufacture.

Table: *Scenarios for modelling land management practices, Santa Maria.*

Scenario	*Land management*
1	Conventional tillage, inorganic fertilizer
2	No-till, inorganic fertilizer
3	No-till, FYM 1.5 tonnes/ha/year, inorganic fertilizer
4	No-till, FYM 3.3 tonnes/ha/crop, no inorganic fertilizer
5	No-till, green manure 10 tonnes/ha/crop, inorganic fertilizer
6	No-till, FYM 1.5 tonnes/ha/year, green manure 10 tonnes/ha/crop, no inorganic fertilizer

Summary

Both models suggest that adopting no-tillage will halt the decline in soil C. However, to increase CS, higher organic additions are necessary (green manures and FYM), which can be used to replace the inorganic fertilizer applications.

Cordoba Province, Buenos Aires Province, and La Pampa Province

Following an equilibrium period with natural vegetation and subsequent prairie conditions from the mid-1800s, CENTURY was parametized with a cultivation regime commencing in the 1950s. This included a fouryear cropping (wheat - soybean - maize - soybean) and a four-year prairie cycle with inorganic fertilizer applications (100 kg/ha urea) starting in 1985. In 1987, a rotated and a nonrotated cropping system was applied that is similar to cultivation practices occurring in the field. The rotated crop sequence was winter forage - soybean - maize - soybean - wheat - soybean - maize - four years of prairie and winter forage - wheat - soybean - maize - soybean - maize - prairie. The non-rotated crop sequence was similar but without the prairie interludes.

Modelled results for the nonrotated crop system show an initial steep fall in soil C to 37 tonnes/ha in 2000 and further losses over subsequent years. RothC estimates a proportionately larger fall from a higher base. The rotated crop - prairie system does not show the same sharp decline in soil C as the nonrotated system did and oscillates around a level of slightly more than 40 tonnes/ha. However, RothC estimates a slightly larger proportionate decline in C for the rotated crop system. Both models calculate smaller differences between the rotated and non-rotated cropping systems in 2000 compared with the difference of 8.5 tonnes/ha measured in the field.

Table: *Total soil carbon for rotated and non-rotated plots, modelled with CENTURY and RothC.*

*Scenario**	*CENTURY*			*RothC*		
	2000	*2050*	*% change*	*2000*	*2050*	*% change*
	(tonnes/ha)			*(tonnes/ha)*		
Non-rotated plots	37.22	32.74	-12.0	50.61	41.17	-18.7
Rotated plots	42.47	40.91	-3.7	54.62	49.53	-9.3
4		45.54	7.2		62.86	15.1

Table: *Scenarios for modelling land management practices, Santa Maria.*

Scenario	*Land management*
1	Non-rotated plots, inorganic fertilizer
2	Rotated plots, inorganic fertilizer
3	Rotated plots, no-till, inorganic fertilizer
4	Rotated plots, no-till, FYM 1.5 tonnes/ha/year, green manure 10 tonnes/ha/crop, no inorganic fertilizer
5	No-till, FYM 1.5 tonnes/ha/year, inorganic fertilizer
6	No-till, green manure 10 tonnes/ha/crop, inorganic fertilizer
7	No-till, FYM 3.3 tonnes/ha/crop, no inorganic fertilizer

Effect of no-tillage

The adoption of a no-tillage regime for the rotated-plots system increases soil C by 2.5 tonnes/ha in the next 50 years, representing a CS rate of 0.02 tonnes/ha/year. The rate increases to about 0.1 tonnes C/ ha/year if green manures and FYM are used instead of fertilizers.

Effect of Organic Inputs

Additions of green manure and FYM with or without inorganic fertilizer can lead to CS rates of 0.06 - 0.13 tonnes/ha. Organic material can replace inorganic fertilizer successfully.

Summary

The inclusion of prairie interludes in the cropping system is an important factor for reducing the decline in soil C. However, the models show that no-tillage and organic inputs are required if C is to be sequestered in this system.

Conclusions from Argentina Cases

The modelling of farm data from three dryland provinces of Argentina shows that carbon stocks have fallen substantially since the prairies were opened up for cultivation. At all three locations, there have been sharp falls in soil carbon stocks, with losses of about 15 tonnes/ha. However, the adoption of no-tillage systems in recent years has halted these declines and, on their own, resulted in small annual increases in soil C of the order of 0.02 tonnes/ ha/year. Rotations with significant periods for return to prairie grassland (e.g. 4 years in 11) result in further increases in soil C. The highest rates of sequestration (0.1 - 0.25 tonnes/ha/year) occur when no-tillage systems also include cultivation of green manures and additions of FYM.

Case Study 5 - Senegal - Old Peanut Basin

The Senegal study area, the "Old Peanut Basin", is located in the west-central part of the country. The climate is semi-arid with annual precipitation ranging of 350 - 700 mm. Almost all arable land is used for rainfed agriculture, with millet, groundnuts, sorghum and cowpea as major crops. The rainy season usually lasts from July to September/ October. However, both spatial and temporal variation in rainfall are high and episodic crop failures are not uncommon. The natural vegetation, including *Faidherbia albida* and various other tree species, has become heavily degraded, primarily because of a long agricultural history and increasing population pressure.

As described in Tschakert (2004b), CENTURY model simulations suggest that soil C in the study area decreased from 20.1 tonnes C/ ha under a native savannah environment in 1851 to 11.9 tonnes C/ ha in 2001. This indicates an annual loss of soil C of 0.055 tonnes C/ ha/year. Tree C declined from 33.6 tonnes/ha to 4.2 tonnes/ha, corresponding to an annual decrease of 0.2 tonnes C/ha/year.

Under improved management conditions (assuming a 50 - year period), soil C could increase by 0.3 - 13.5 tonnes/ha, or 0.006 - 0.27 tonnes/ha/year. C gains in trees could be tripled (from 4.2 to 11.8 tonnes/ha), assuming a conversion of croplands to grassland - tree plantations. Given the fact that most of the C gains are achieved in the first 25 years, annual increases for this time period range from 0.02 to 0.43 tonnes C/ha/year, which is higher than the estimates provided by Lal, Hassan and Dumanski (1999). Under poor management, in this case an annual millet - sorghum rotation with no inputs and permanent browsing and pruning of tree resources, both soil and tree C continue to drop, reaching an absolute minimum level of 7.9 tonnes/ha and 0.6 tonnes/ ha, respectively.

The various improved management practices with the anticipated changes in soil C over two time periods (2002 - 2026 and 2027 - 2050) as discussed by Tschakert (2004b). As illustrated, considerably higher gains in soil C can be achieved in the first 25 years, except for the tree-plantation scenario. However, in some cases, these gains cannot be sustained in the second twenty 25-year period, and losses would have to be expected if no additional inputs occurred. This is particularly true for the fallow scenarios.

Table: *Effects of land management practices or land use on carbon sequestration potential in the Old Peanut Basin, Senegal*

Technological options*	***Change in soil C (tonnes/ha/year) 2001 - 2026 (first 25 years)***	***Change in soil C (tonnes/ha/year) 2027 - 2050 (second 25 years)***
Compost (2t)	0.02	-0.01
Conversion of croplands to grasslands + grazing	0.06	0.02
3-year fallow + 2 tonnes manure in rotation with 4 years cropping	0.14	-0.05
Cattle manure (4 tonnes)	0.10	0.01
Conversion of croplands to grasslands	0.17	0.04
Cattle manure (4 tonnes) + min. fertilizer (250kg on millet; 150kg on	0.12	0.01
groundnuts)		
Sheep manure (5 tonnes)	0.13	0.01
3-year fallow + Leucaena prunings (2 tonnes) in rotation with 4 years	0.18	-0.05
cropping		
Conversion of croplands to grasslands with tree protection	0.10	0.04
Sheep manure (10 tonnes)	0.17	0.01
10-year fallow + 2 tonnes manure in rotation with 6 years cropping	0.25	-0.04
10-year fallow + Leucaena prunings (2 tonnes) in rotation with 6 years cropping	0.25	-0.04
Conversion of croplands to grasslands + tree plantation (*Faidherbia albida*)	0.23	0.21
Agricultural intensification (improved millet, manure, Leucaena prunings, min. fertilizer, animal traction, 1-year fallow)	0.43	0.11

Agricultural intensification (improved millet, manure, Leucaena prunings, min. fertilizer, animal traction, 1-year fallow) 0.43 0.11

*All technological options calculated for 1 ha. Annual rotation between millet and groundnuts for cropping scenarios.

Source: Tschakert (2004b).

Case Study 6 - Sudan - Northern Kordofan Province

The study was undertaken in Northern Kordofan Province in the Sudan, an area dominated by coarse-textured soils of Aeolian origin, locally named *Qoz*. The study site is representative in terms of soils, climate and vegetation type for a large region stretching from the Atlantic coast to the Ethiopian highlands (Olsson and Ardö, 2002; Olsson and Tschakert, 2002).

The climate is semi-arid with annual rainfall ranging from less than 200 mm in the north to about 350 mm in the south. Land use can be characterized as a gradual increase in intensity with rainfall. In the northern part, only very extensive grazing by camels is possible owing to the lack of permanent water sources. Cultivation of mainly millet and to a lesser extent sesame and groundnuts predominates where rainfall and water resources increase towards the south. Grazing by cattle, sheep and goats is also an important factor in the cultivated parts of the area.

Interviews and literature indicate that land-use practices have changed markedly from a rotation system with long fallow periods (15 - 20 years), interspersed with short periods of cultivation (4 - 5 years), to more continuous cultivation in the last three to four decades. In the same period, crop yields have decreased, mainly because of a marked decline of rainfall but to some extent also due to the abandonment of fallow periods. Increased demand for food (as a result of population increase) combined with decreasing yields has forced farmers to extend their cultivated area primarily by reducing fallow periods.

In order to investigate the potential for increasing the soil carbon content through land-management changes, a number of future scenarios were simulated. First, the equilibrium level was established by running the model for 2 000 years with no land use, and then six different land-management scenarios were simulated for the period 2000 - 2100. The land-management scenarios were: continuous cultivation, crop-tofallow ratios of 5:6, 5:10, 5:15, 5:20, and no cultivation but only grazing.

While the results come from model simulations, the results show empirical research, using soil sampling, linking fallow periods to increased levels of SOC. In order to investigate the role of the length of fallow periods, fields in different stages of the crop rotation were sampled and their corresponding SOC levels were determined (Olsson and Ardö, 2002). The graph verifies to a large extent the model results and shows that there is significant potential for increasing SOC even in these dry sandy soils.

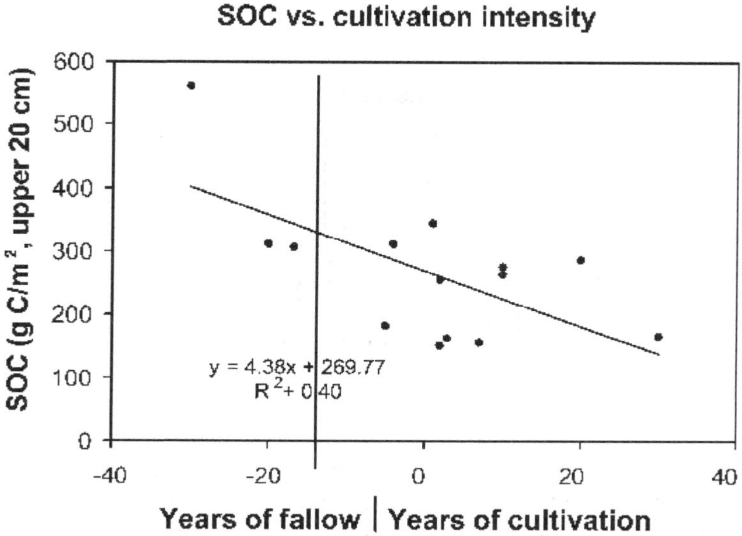

Figure: *SOC in relation to fallow and cultivation history in fields in Northern Kordofan Province in the Sudan*

Source: Olsson and Ardö (2002).

Figure shows that increasing the length of the fallow periods, i.e. decreasing the crop - fallow ratio, causes an increase in the soil carbon content proportional to the ratio (i.e. longer fallow more soil C). A landuse change from millet cultivation to grazing is estimated to increase the soil carbon content from about 1.5 tonnes/ha to 2.5 tonnes/ha in 100 years. This is 82 percent of the equilibrium phase prior to millet cultivation. If millet cultivation is continued into the future, a further decrease in the soil carbon content is expected to 0.68 tonnes/ha in 2100. Changing the continuous cultivation to crop - fallow ratios of 5:6, 5:10, 5:15 and 5:20 will increase the soil carbon content.

5

Salinity Problems of the Dryland Regions

Apart from irrigated areas, salinity poses a major management problem in many unirrigated areas where cropping is done under rainfed conditions. Dryland salinity has been a threat to the land and water resources in several parts of the world although only in recent years has the seriousness of the problem become widely known. Dryland salinity is an acute management problem in western Australia and in the Great Plains region of North America. In Canada it occurs extensively in the prairie provinces of Manitoba, Saskatchewan and Alberta and in the United States in the states of Montana, North and South Dakota. Dryland salinity is also said to occur in South Africa, Iran, Afghanistan, Thailand and India and it probably exists in other countries. Saline spots or areas occurring in the dryland fields have been known by several local names, but most commonly as saline seeps. These problem soils range from a slightly saline soil condition which reduces crop growth to extensive areas where cultivation is almost impossible. Problems relating to the origin and management of dryland saline soils have been discussed at several meetings in the recent past (Anonymous 1976, 1978; Holmes and Talsma 1981).

Mode of Formation

Dryland saline seepage is generally considered to be a manifestation of salt accumulation in seepage spots at low points or side slopes in the landscape developed when water infiltrates into soil to somewhat impermeable layers, and moves laterally downslope. Eventually at lower elevations, the water seeps laterally out at the soil surface and evaporates leaving the salts behind. Thus the development of saline seeps involves two areas in the field - the recharge and the discharge areas. In the recharge areas, water in excess of the retention capacity of the root zone soil percolates beyond the root zone, reaches the groundwater and increases the flow to the discharge area. The groundwater flow is mainly lateral and downslope and occurs most often over a shallow, less permeable layer. The groundwater travelling to the discharge area dissolves salts from the soil. In the discharge area the groundwater rises to the soil surface creating a seep. As the water evaporates from the seepage area salts accumulate.

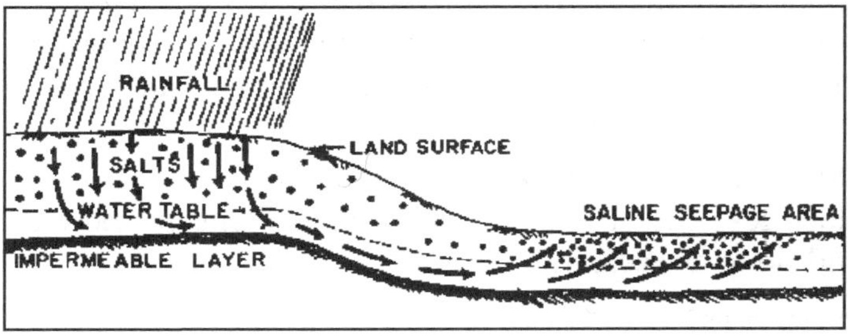

Figure: *Schematic diagram of a recharge and a seepage area (zone of salt accumulation).*

Factors Influencing Formation of Saline Seeps

A combination of climatic, geologic, soil and cultural conditions determines the nature, extent and intensity of saline seeps.

Excess Water

Many dryland areas which are now cropped, existed under forest or grass cover. The ecosystem under these conditions was balanced - the grasses and trees utilizing all the precipitation in their respective areas and keeping the groundwater tables low. Clearing deep rooted perennial vegetation and its replacement by pastoral species and crops decreased plant water requirements, causing a surplus exceeding the retention capacity of the root zone. Any land use practice which allows excess moisture to migrate downward through the soil profile beneath the root zone can contribute to the formation and extension of dryland salinity. Some of the more important factors and practices that encourage excess moisture are:

i. ***Fallowing:*** By far the most important contributing factor in many areas, e.g. Northern Plains of North America is the widespread alternate crop-fallow farming system. When fallowing is practised, the capacity of the soil to retain rain water is reduced resulting in greater losses through seepage.

ii. ***Rainfall:*** It has generally been observed that salinity problems are more severe in years of high rainfall due to greater recharge and therefore accumulation of salt-laden water in the recharge zone.

iii. ***Water and snow accumulation:*** Practices which result in the accumulation of water/snow in the recharge areas may aggravate the saline seep problem.

These include shelter belts, single row wind breaks, railways, roads and highways, etc. Natural and constructed ponds, dugouts and small farm reservoirs can be important locally if they are located in the recharge areas and leak significant quantities of water.

iv. ***Overgrazing:*** Range that is overgrazed utilizes and holds in place less moisture than before resulting in increased runoff and deep percolation.

Soil Characteristics

Light textured soils having low moisture retention and high permeability in the recharge zone are more conducive to the formation of saline seeps at low points in the landscape.

In some places the bed rock is of marine origin and its overburden is rich in soluble salts. Dissolution of this salt and its transportation with percolating water causes serious salinity problems.

In many areas having saline seep problems the bed rock or the overburden are not of marine origin but the salts are derived from the normal weathering of the mineral constituents of the soil. Excess salts are rarely a problem in the place of their origin.

Prevention and Management of Saline seeps

Since the cause of saline seeps is nearly always increased recharge in an upper area, any long-term solutions to the problem must include regional land use changes with the objective of at least partly restoring the original hydrological state.

Apart from measures to restore the hydrologic equilibrium, site specific treatments of salt-affected land are required to restore their productivity.

Since the problem is invariably complex, it is essential to devise an effective strategy for each region involving both regional land use changes and site treatment. For such strategies detailed investigations of the processes, sources and movement of salts must be worked out. Some of the more frequently recommended practices are discussed below.

Practices for the Recharge Area

Any practices that will increase water use in the recharge areas or in other ways decrease the excess water which ultimately contributes to the seepage will help in the control of saline seeps.

i. Intensive cropping Intensive cropping and elimination of long fallows result in increased water use in the recharge area, decreased seepage flow and therefore a reduced salinity problem.

ii. Growing deep rooted perennials Crops differ significantly in their rooting depth and seasonal water use. Research in Montana (Brown and Miller, 1978) showed that among the crops tested, alfalfa had the deepest rooting system (6.10 m) and depleted the maximum soil water (787 mm).

   These scientists studied the effect of six years of cropping with alfalfa on a saline seep recharge discharge area. In 1971, the watertable depth was 0.30 m below the soil surface in the discharge area and at 6.2 m in the recharge area.

   Six years later the watertable dropped by 2.9 m in the discharge area and by 2.0 m in the recharge area. Research is also underway to test several other crops and cropping sequences in relation to control of saline seeps.

iii. Drainage Surface drainage of natural and man-made ponds in the recharge area is an effective, important and often inexpensive method of controlling excess water in the seepage areas.

   Drainage of ponded water in the recharge area should be preferred over drainage measures in the seep area. In the latter, additional problems are encountered which include disposal of saline effluents, lack of adequate outlets, etc.

   While surface drainage may be adequate for removing excess water from the recharge areas, it is often necessary to resort to subsurface drainage in the discharge areas which is a costly proposition.

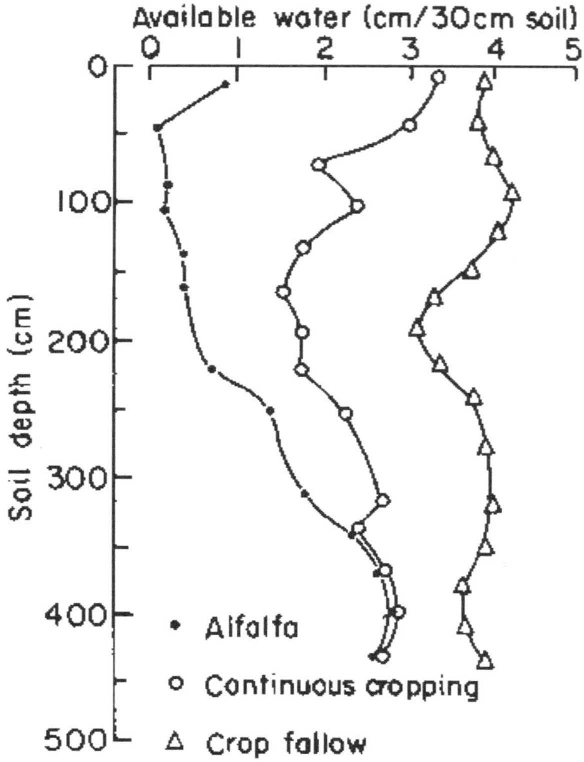

Figure: *Plant available soil water content as influenced by various cropping systems (Brown et al., 1976)*

Practices for the Discharge Areas

i. ***Cropping:*** As in the case of recharge areas, intensive cropping involving perennial and deep rooted crops having a high water use helps in maintaining the watertable at greater depths and checks salinization.

ii. ***Drainage:*** The objective of drainage in the discharge areas is to lower the watertable and desalinize the root zone through leaching. Artificial subsurface drainage and leaching by applied water are generally not economic in the dryland areas. Even so, drainage works have been installed with variable performance in limited areas. Sommerfeldt et al. (1978) experimented with mole drains which are made by pulling a bullet shaped object through the soil with a mole plough. It consists primarily of a beam with a perpendicular blade extending downwards. On the end of the blade is the bullet shaped object which leaves a channel. The drains are quickly installed and the installation cost is low. Mole drains perform best in soils with sufficient clay content to be cohesive and where the soil is wet enough to be plastic. The disadvantages of the mole drains are the limited depth (60 to 70 cm) to which these can be made and their relatively short life span (2 to 3 years).

According to Van der Pluym (1978) and Miller et al. (1981) subsurface drainage of most saline seep discharge areas is unfeasible because of extremely low permeability of soils resulting from high content of montmorillonite clay and often high SAR of the soil solution, problems of waste disposal, poor accessibility for heavy drainage equipment and difficulty in locating saline seep focal points. Also the cost of materials and installation is relatively high and is often more than the land price. From these considerations, it would be better to provide drainage and adopt excess water control measures in the recharge area rather than to control the watertable through drainage in the discharge areas.

iii. ***Salt-tolerant crops:*** Planting salt-tolerant species is an effective way of obtaining some economic returns while efforts are being made to improve the saline seeps. Investigations on the relative tolerance of plant species in several Australian states have indicated the value of growing the following in the saline seep areas (Mitchel et al. 1978):

Agropyron elongatum	Puccinellia ciliata
Atriplex spp.	Sporobolus virginicus
Hordeum hystrix	Trifolium fragiferum;
Kochia brevifolia	T. glomeratum
Lolium rigidum	T. resupinatum
Paspalum vaginatum	T. subterranum

Studies in Canada (Van der Pluym, 1978) showed that for soils affected by moderate salinity levels, ECe up to 8.5 dS/m, among the cereal crops tested, 6-row barley outyielded all other types and varieties. For soils with higher salinity levels a mixture of alfalfa and salt-tolerant grasses like Agropyron elongatum, Agropyron cristatum, Agropyron trachycaulum and Festuca elatior was found to perform well. Where soil salinity levels were even higher, (more than 14.0 dS/m), annual crops or grasses were not recommended. Under these conditions Kochia scoparia, a prolific growing native weed, was allowed to grow and had excellent feed value.

Soil and Water Relationships

Soil moisture limits forage production potential the most in semiarid regions. Estimated water use efficiency for irrigated and dry-land crop production systems is 50 percent, and available soil water has a large impact on management decisions producers make throughout the year. Soil moisture available for plant growth makes up approximately 0.01 percent of the world's stored water.

By understanding a little about the soil's physical properties and its relationship to soil moisture, you can make better soil-management decisions. Soil texture and structure greatly influence water infiltration, permeability, and water-holding capacity.

Soil texture refers to the composition of the soil in terms of the proportion of small, medium, and large particles (clay, silt, and sand, respectively) in a specific soil mass. For example, a coarse soil is a sand or loamy sand, a medium soil is a loam, silt loam, or silt, and a fine soil is a sandy clay, silty clay, or clay.

Soil structure refers to the arrangement of soil particles (sand, silt, and clay) into stable units called aggregates, which give soil its structure. Aggregates can be loose and friable, or they can form distinct, uniform patterns. For example, granular structure is loose and friable, blocky structure is six-sided and can have angled or rounded sides, and platelike structure is layered and may indicate compaction problems. Soil porosity refers to the space between soil particles, which consists of various amounts of water and air. Porosity depends on both soil texture and structure. For example, a fine soil has smaller but more numerous pores than a coarse soil. A coarse soil has bigger particles than a fine soil, but it has less porosity, or overall pore space. Water can be held tighter in small pores than in large ones, so fine soils can hold more water than coarse soils.

Water infiltration is the movement of water from the soil surface into the soil profile. Soil texture, soil structure, and slope have the largest impact on infiltration rate. Water moves by gravity into the open pore spaces in the soil, and the size of the soil particles and their spacing determines how much water can flow in. Wide pore spacing at the soil surface increases the rate of water infiltration, so coarse soils have a higher infiltration rate than fine soils.

Permeability refers to the movement of air and water through the soil, which is important because it affects the supply of root-zone air, moisture, and nutrients available for plant uptake. A soil's permeability is determined by the relative rate of moisture and air movement through the most restrictive layer within the upper 40 inches of the effective root zone. Water and air rapidly permeate coarse soils with granular subsoils, which tend to be loose when moist and don't restrict water or air movement. Slow permeability is characteristic of a moderately fine subsoil with angular to subangular blocky structure. It is firm when moist and hard when dry.

Water-holding capacity is controlled primarily by soil texture and organic matter. Soils with smaller particles (silt and clay) have a larger surface area than those with larger sand particles, and a large surface area allows a soil to hold more water. In other words, a soil with a high percentage of silt and clay particles, which describes fine soil, has a higher water-holding capacity. The table illustrates water-holding-capacity differences as influenced by texture. Organic matter percentage also influences water-holding capacity. As the percentage increases, the water-holding capacity increases because of the affinity organic matter has for water.

Water availability is illustrated in the figure by water levels in three different soil types. Excess or gravitational water drains quickly from the soil after a heavy rain because of gravitational forces (saturation point to field capacity). Plants may use small amounts of this water before it moves out of the root zone. Available water is retained in the soil after the excess has drained (field capacity to wilting point). This water is the most important for crop or forage production. Plants can use approximately 50 percent of it without exhibiting stress, but if less than 50 percent is available, drought stress can result. Unavailable water is soil moisture that is held so tightly by the soil that it cannot be extracted by the plant. Water remains in the soil even below plants' wilting point.

One can see from the table that soil texture greatly influences water availability. The sandy soil can quickly be recharged with soil moisture but is unable to hold as much water as the soils with heavier textures. As texture becomes heavier, the wilting point increases because fine soils with narrow pore spacing hold water more tightly than soils with wide pore spacing.

Soil is a valuable resource that supports plant life, and water is an essential component of this system. Management decisions concerning types of crops to plant, plant populations, irrigation scheduling, and the amount of nitrogen fertilizer to apply depend on the amount of moisture that is available to the crop throughout the growing season. By understanding some physical characteristics of the soil, you can better define the strengths and weaknesses of different soil types.

Soil Water (Retention)

Soils can process and contain considerable amounts of water. They can take in water, and will keep doing so until they are full, or the rate at which they can transmit water into, and through, the pores is exceeded. Some of this water will steadily drain through the soil (via gravity) and end up in the waterways and streams. But much of it will be retained, away from the influence of gravity, for use of plants and other organisms to contribute to land productivity and soil health.

Soil Water Retention

The spaces that exist between soil particles, called pores, provide for the passage and/or retention of gasses and moisture within the soil profile. The soil's ability to retain water is strongly related to particle size; water molecules hold more tightly to the fine particles of a clay soil than to coarser particles of a sandy soil, so clays generally retain more water (Leeper and Uren, 1993). Conversely, sands provide easier passage or transmission of water through the profile. Clay type, organic content and soil structure also influence soil water retention (Charman & Murphy 1977). The maximum amount of water that a given soil can retain is called field capacity, whereas a soil so dry that plants cannot liberate the remaining moisture from the soil particles is said to be at wilting point (Leeper & Uren 1993). Available water is that which the plants can utilise from the soil within the range of field capacity and wilting point. The role of soil water retention is profound - its effects are far reaching and relationships are invariably complex.

Soil Water Retention and Organisms

Soil water retention is essential to life. It provides an ongoing supply of water to plants between periods of replenishment (infiltration) so as to allow their continued growth and survival. Over much of temperate Victoria, Australia, for example, this effect is seasonal and even inter-annual; the retained soil water that has accumulated in preceding wet winters permits survival of most perennial plants over typically dry summers when monthly evaporation exceeds rainfall.soils generally contain more nutrients,moisture and humus.

Soil Water Retention and Climate

Soil moisture has an effect on the thermal properties of a soil profile, including conductance and heat capacity, explains Oke (1987); the association of soil moisture and soil thermal properties has a significant effect on temperature-related biological triggers, including seed germination, flowering and faunal activity.

Recent climate modelling by Timbal et al. (2002) suggests a strong linkage between soil moisture and the persistence and variability of surface temperature and precipitation; further, that soil moisture is a significant consideration for the accuracy of "inter-annular" predications regarding the Australian climate.

Soil Water Retention, Water Balance and Other Influences

The role of soil in retaining water is significant in terms of the hydrological cycle; including the relative ability of soil to hold moisture and changes in soil moisture over time:

- Soil water that is not retained or used by plants may continue downward through the profile and contribute to the water table, the permanently saturated zone at the base of the profile – this is termed recharge. Soil that is at field capacity (among other reasons) may preclude infiltration so to increase overland flow. Both effects are associated with ground and surface water supplies, erosion and salinity.
- Soil water can affect the structural integrity or coherence of a soil – saturated soils can become unstable and result in structural failure and mass movement. Soil water, its changes over time and management are of interest to geo-technicians and soil conservationists with an interest in maintaining soil stability.

Conclusion

Soil water retention is widely studied and reported on in the literature due to its extensive and profound role; it is a key part of the hydrological cycle, provides support to organisms, interacts with climate, and is a major consideration in ground and surface water supply, environmental and geo-technical aspects.

Agriculture and Dry Lands

Dry lands present challenges for agriculture because plants require soil with adequate moisture to grow in healthy and productive ways. When a plant's water requirements are greater than the soil can provide then they will not yield.

Before considering the problem of agriculture and dry lands it is important to know the history of the land. It could be traditionally dry or it may only be a case of just a few decades that dry lands have occurred due to processes that are destructive to the water cycle such as deforestation.

Humans use water for agriculture more than any other purpose and according to the United Nations it claims 70% of all freshwater appropriated for human use. With food demand predicted to increase 50% by 2030 demand for water will certainly increase. With increased demand water use in arid regions must become more efficient and even attempt to create a climate with more moisture through processes such as re-forestation and water harvesting.

Moisture content in dry lands is not only concerned with how much rainfall an area receives or its proximity to major water sources but also factors how much moisture is retained by soil and the plants living within that soil. A big factor is the number of plants within the region that are able to hold onto moisture. If the area was heavily deforested , only promotes the growth of one kind of plant in a monoculture, or the combination of the two then the moisture content of the soil will be seriously compromised.

Agriculture production systems in dry areas are often over-stretched. Solutions to the problems created by dry lands need to be simple for the farmer to implement and accessible enough for them to empower themselves on their own. While applications such as strategic policy, capacity building, and technology are creating good solutions farmers need their own hands on, simple solutions that they can implement themselves and create productive dry lands that will retain moisture in healthy soil for now and for future generations. Solutions such as re-forestation, mulching, and building water retention with increased and diverse plant life are all important.

The Prevailing Context

Major dry lands are located in Asia, Africa and Latin America. Many of these areas are characterized by environmental challenges such as variable precipitation and temperatures, economic and physical water scarcity, low soil fertility, severe land degradation and loss of biodiversity. To make the situation even more complex, dry lands are generally economically and politically marginalized, and struggle with socioeconomic constraints such as lack of investment and infrastructure. National and international investment strategies tend to favour high potential or densely populated areas. Many small land holder systems in dry areas have poor access to markets and to inputs such as improved seeds and seedlings, fertilizers, livestock, animal services, technology and information about alternative production systems.

Solutions and Related Challenges

Low-Tech Solutions: Farmers need to be empowered to handle the challenges that dry lands provide. Whether or not the dry lands were man-made there are solutions that require only simple techniques to be applied in order to realize major increases in the moisture content of soil.

Reforestation

In many parts of the world dry lands have been artificially created through irresponsible deforestation practices. It is important for the dry land farmer to understand the history of his land and whether or not it was once rich in trees or other plants. Through the act of deforestation humans have created a significant loss of moisture throughout agricultural lands and consequently plant health that depends so heavily on moisture suffers greatly.

Deforestation not only causes a loss of rainfall in agricultural lands it also significantly reduces moisture in deforested areas. Permaculture pioneer Bill Mollison asserts that forests are crucial for precipitation and that precipitation can decrease by over 86% in deforested areas due to loss of rainfall and more significantly loss of moisture retention. Furthermore, dry lands which are devoid of trees through human actions carry soils which have a high content of salinity. The cause of this again is directly related to deforestation where cutting trees causes loss of clean water down the water table. Trees act as pumps to keep salt low in the table and fresh water high. When the forest is cut the salt rises and will in turn degrade the soil in a short period of time.

Small scale re-forestation needs to be a priority of farmers for these reasons. The short immediate gains provided by deforestation are strongly outweighed by the long term devastation caused by removing trees.

Recovery Through Plant Growth

Plants hold major amounts of moisture. In order to start the process of moisture retention in an arid region plants must be grown that are drought resistant and can handle prolonged periods without water. After successfully creating a diverse area of drought resistant plants the moisture content of the area will be increased and allow for plants to be grown that have larger water requirements. Here we can take advantage of what Mollison explains is the greatest reservoir we have on a farm, the soil. This is a process that can take decades to properly implement but the payoff will be immense for future generations working the land.

Mulch

Soils in agriculture lose much moisture through direct contact with the sun when moisture within the soil is evaporated through direct sunlight. Placing a layer of mulch on top of soil such as dry grass or leaves will retain much of the moisture within soil and even with weeks of sustained sunlight. Since the soil is not exposed directly it will keep soil shady, rich in moisture, and healthy. Plants can be grown by removing small spaces of mulch to let them grow. Repeating the process as the mulch decomposes is crucial.

Technological Solutions

Modern Irrigation Technologies

One of the main causes for the limited productive use of water resources in dry regions is limited access to modern irrigation technologies. In areas with limited water availability, access to adequate irrigation is responsible for abut 70% of the increase in crop yields. For example, drip irrigation is a water-saving technology which enables slow and regular application of water directly to the roots of the plants through a network of economically designed plastic pipes and low discharge emitters. It maximizes crop productivity through increase in the crop yield and also the area for cultivation, and protects the environment through conserving soil, water and fertilizer resources, thus increasing the farmer income. Currently, this is being promoted by the Government of India in the form of kits in the water scarce regions in India. These include: Drum Kit; Bucket Kit; Family Nutrition Kit and Customised Systems.

Biotechnology

According to the United Nations Conventions on Biological Diversity, it is a "technological application that uses biological systems, living organisms, or derivatives thereof, to make or modify products or processes for specific use". Today genetic engineering is used widely for the following purposes:improve yield from crops; reducevulnerability of crops to environmental stresses; increase nutritional qualities of food crops; improve taste, texture or appearance of food; reduce dependence on fertilizers, pesticides and other agrochemicals; produce novel substances in crop plants. An example of one type of biotechnology application are drought resistant seeds. The Story of Mutubanda - drought and climate change in Sri Lanka (by PracticalAction) describes how one farmer uses traditional drought-resistant seed varieties to adapt to changing climate conditions.

Strategic Public Policy

Policies at a country level need to be thought as a medium to long term measurese to enhance income generating options for the rural poor from crops and livestock, especially small ruminants, by improving, intensifying and diversifying current agricultural production systems, sustaining the resource production base, increasing and diversify outputs, improving safety, quality and marketability of produce and adding value through agri-processing of primary products. Policies should aim to contribute to the development in rural areas of productive and sustainable systems that enhance livelihoods, generate opportunities for rural agri-business development and increased employment and conserve the resource base. For dryland areas, it is necessary to have a new development paradigm to move dryland towards a more parsimonious set of systems-oriented principles that can help foster local economic growth even under current climate change scenerios. The key areas of such paradigm shall be based on key development areas such as reducing rural poverty, improving food security, improving nutrition and health and sustainable management of natural resources. these lines shall be linked to on-going or future policy programmes of countries in Asia and Africa that have dryland development among the core problems (source: CGIAR,2011).

Capacity Building

Some capacity building elements to considering when developing adequate agricultural production systems include, but are not limited to:

1. Water management and accounting
2. Adaptation strategies for farming and for local communities
3. Technical training for farmers

Design of Cropping Systems to Achieve Nutritional Goals in Dry Farming

During the past three decades India has made rapid strides with regard to food-grain production, from about 50 million tonnes in 1950/51 to 132 million tonnes in 1980/81. The increase in irrigated area, from 27 million hectares to 60 million hectares, on the one hand, and introduction and development of high yielding varieties and hybrids of cereals, on the other, contributed to this increase. However, the production of pulses and oilseeds remained almost static during this period. Despite the fact that rainfed lands in India contribute to more than 90 per cent of the total production of pulses and oilseeds, the dryland farmers inhabiting these lands, continue to be on the low plane of calorie and protein nutrition.

Research results from the Dryland Agricultural Research Project (All-India Coordinated Research Project for Dry/and Agriculture) show that through intercrop and double crop systems, production of pulses and oilseeds can be increased without unduly sacrificing the yields of cereals and millets. Cropping systems are now available that can meet the nutritional goals {about 10,000 calories per day for an average Indian family of five members) in dry farming areas.

Among the cereal based intercropping systems, maize + soybean supplies the highest calories and other nutrients on a per hectare basis. However, the sorghum + pigeon-pea (2:1 row ratio) system is more stab/e over the seasons and environments and, hence, preferable. Among sequential cropping systems. the sorghum-chick-pea system holds promise of providing better nutritional standards. From the point of view of biological value and protein efficiency ratio, the rice-chick-pea sequence cropping system holds the greatest promise.

Drylands in India contribute about 75 per cent of the cultivated acreage. Even after the full irrigation potential of the county is achieved in due course, more than 50 per cent of the acreage will continue to be rain-fed. Rain-fed lands will, therefore, remain the major source for the supply of millets, pulses, and oilseeds to the growing population. Currently, these lands contribute to 80 per cent of the total production of millets and more than 90 per cent of total production of pulses and oilseeds in the country

In India, as in many Asian and African countries, the nutritional problem is one of undernutrition rather than malnutrition. It is inadequate food intake, coupled with the predominance of cereals/ millets in the diet, that lead to undernutrition. Several reasons are given for the inadequate consumption of food by several population subgroups in India. Of these, lack of purchasing power, low per capita food production and availability, and large deficits in production of pulses and oilseeds that are rich sources of calories, proteins and fats, are important. Milk and meat, known for their high calorie, protein and fat contents, besides being costly, are in short supply.

Increasing and stabilizing the production of cereals/millets on the one hand, and introducing more pulses and oilseeds in the cropping systems, on the other, will go a long way towards meeting the nutritional goals of dry farming. Concerted research efforts in the last decade 11970-1980), by multidisciplinary teams engaged in dry farming research at 23 regional research stations in the country, have led to the evolution of cropping systems that hold promise, not only for high productivity, but for meeting the nutritional goals of the dry regions as well.

The Nutritional Goals

Nutritional goals are better appreciated on a family basis than a regional one. A farmer's family, consisting of an adult male, an adult female, an adolescent boy or a girl and two children, require a daily average of 10,600 calories, 340 9 protein, 5.0 9 calcium, and 110 mg iron. On a body-weight basis children require twice as much protein, calcium, and iron as do adults.

Table: *Nutritional Requirement per Day of an Average Farmer's Family*

Household member	*No. of members*	*ACU*[a]	*Calories*	*Protein (g)*	*Calcium (g)*	*Iron (mg)*
Adult male	1	1.0	2,800	55	1.0	25
Adult female	1	0.9	2,300	45	1 0	25
Adolescent (boy/girl)	1	1.0	2 500	80	1 0	20
Children	2	1 4	3,000	160	2 0	40
Total	5	4.3	10,600	340	5.0	110

a. Actual Consumption Unit.

Cropping Systems for Drylands and Their Production Potential

Cereal-based Systems

Sorghum, pearl millet, and finger millet are the major millets grown in dry regions. In regions of Telengana, Marathwada, Vidarbha, the Malwa plateau and parts of Gujarat - areas of more than 800 mm rainfall - the intercrop system of sorghum + pigeon-pea (2:1) has shown promise, with a production potential of 2.0 to 2.5 tonnes per hectare (t/ha) of sorghum grains and 0.6 to 0.8 t/ha of pigeon-pea grains, against a sole crop of either sorghum (2.5 t/ha) or pigeon-pea (0.8 t/ha). In the black soil regions of Andhra Pradesh, Karnataka, Maharashtra, and Madhya Pradesh where the rainfall is more than 1,000 mm/annum, the double crop system with sorghum, followed by chick-peas, has been found to do exceedingly well. The system is capable of giving around 3.0 to 3.5 t/ha of total productivity, in terms of grains.

Table: *Sorghum-based Cropping Systems for Drylands and Their Production Potential in a Normal Season*

Cropping system	*Region(s)*	*Yield (t/ha) of sorghum*		*Yield (t/ha) of pigeon-pea/chick-pea*	
		Grain	*Edible portion*	*Grain*	*Edible portion*
Intercrop system					
Sorghum + pigeon-pea(2:1)	Telengana, Marathwada, Vidharbha, Malwa plateau, and parts of Gujarat, where rainfall > 800 mm	2.0- 2.2	2.0- 2.2	06- 0.8	0.5- 0.6
Sorghum (sole)		2.5	2.5	0.8	0.8
Double crop system					
Sorghum - chick-pea	Black soils of Andhra Pradesh, Karnataka, Maharashtra, and Madhya Pradesh, where rainfall > 1,000 mm	2.0 2.5	2.0 2.5	10 (whole)	10 (whole) or 0.6- 0.8 (dhal)

***Table:** Pearl Millet-based Cropping Systems for Drylands and Their Production Potential in a Normal Season*

Cropping system	*Region(s)*	*Yield (t/ha) of pearl millet*		*Yield (t/ha) of green gram/chick-pea/pigeon-pea*	
		Grain	*Edible portion*	*Grain*	*Edible portion*
Intercrop system					
Pearl millet + pigeon-pea	Parts of Telengana; Karnataka, Maharashtra, Gujarat, Madhya Pradesh, Rajasthan, Haryana, and S Tamilnadu	1.2- 1.5	1.0- 1.3	0.6- 0.8	0.5 0.6
Pearl millet + green gram	Jodhpur	1.2- 1.5	1.0 1.3	0.4- 0.6	0.3 0.5
Pearl millet (sole)		2.0	1.7		
Double crop system					
Pearl millet - chick-pea	Varanasi	1.5- 1.6	1.3 1.4	0.8- 1.0	0.8 1.0 (whole) or 0.6-0.8 (dhal)
Pearl millet - mustard	Jodhpur, Hissar	2.0- 2.2	1.7- 1.9	0.8- 1.0	0.2 0.3

In parts of Telengana, Karnataka, Maharashtra, Gujarat, Madhya Pradesh, Rajasthan, Haryana, and south Tamilnadu, which have relatively light soils, the pearl millet + pigeon-pea intercrop system has given a yield potential of 1.2 to 1.5 t/ha of pearl millet grains, and 0.6 to 0.8 t/ha of pigeon-pea grains, against a sole crop of pearl millet (2.0 t/ha).

In the Jodhpur and Hissar regions, which have sierozem soils, with an annual rainfall of 350 to 400 mm, the pearl millet + green gram intercrop system has proved to be best from the point of view of total productivity per unit area.

In seasons of higher rainfall (> 500 mm) with an extended rainy season, the double crop system of pearl millet (BJ-104) followed by mustard (T. 59) can pay rich dividends in areas like Jodhpur and Hissar.

The system is capable of giving a total productivity of about 3 t/ha, of which one-third is contributed by mustard, an oilseed crop.

Yet another intercrop system, having finger millet as the principal crop and soybean as the companion crop, is emerging as a potentially useful system in the Bangalore and Ranchi regions.

***Table:** Finger Millet-based Cropping Systems for Drylands and Their Production Potential in a Normal Season*

Cropping system	Region(s)	Grain yield (t/ha) of finger millet	Grain yield (t/ha) of soybean
Intercrop system			
Finger millet +	Bangalore	1.0-	0.2
soybean	Ranchi	2.0	0.3
Finger millet (sole)		2.5	

***Table:** Maize-based Cropping Systems for Drylands and Their Production Potential in a Normal Season*

Cropping system	Region	Yield (t/ha) of maize		Yield (t/ha) of companion/second crop	
		Grain	Edible portion	Grain	Edible portion
Intercrop system					
Maize + pigeon-pea (1:1)	Bangalore, Akola, Udaipur, Varanasi, Jammu, and Hoshiarpur	2.0- 2.5	2.0- 2.5	0.4- 0.5	0.3 0.4
Maize + soybean	Bangalore, Akola, Indore, and Dehra Dun	2.0- 2.2		1.0 1.2	
Maize + green gram	Udaipur	2.0- 2.2	2.0- 2.2	0.4- 0.6	0.3 0.5
Maize (sole)		2.5	2.5		
Double crop system					
Maize - wheat	Hoshiarpur, Jammu, Dehra Dun	2.5- 2.6		1.8 2.0	
Maize - mustard	Hoshiarpur, Agra, and Udaipur	2.5- 2.6		1.0 1.2	

***Table:** Paddy-based Cropping Systems for Drylands and Their Production Potential in a Normal Season*

Cropping system (sequence cropping)	Region	Yield (t/ha) of paddy		Yield (t/ha) of second crop	
		Grain	Edible portion	Grain	Edible portion
Paddy - chick-pea	Dehra Dun, Varanasi, and Rewa	2.0- 2.5	1.2- 1.5	0.6 0.8	0.5 0.6
Paddy - finger millet	Bhubaneswar, Ranchi, and Rewa	2.0 2.5	1.2- 1.5	0.6- 0.8	0.6 0.8
Paddy - wheat	Dehra Dun, Ranchi, Rewa, and Varanasi	2.0- 2.5	1.2- 1.5	1.8- 2.0	1.8 2.0

Maize-based cropping systems are quite popular in regions like Bangalore, Akola, Udaipur, Varanasi, Indore, Dehra Dun, Jammu, and Hoshiarpur. Intercrop systems of maize + pigeon-pea, maize + soybean and maize + green gram have exhibited distinct superiority in terms of total productivity over the other systems tried. Of these, a maize + soybean intercrop system gave the highest productivity, 2.0 to 2.2 t/ha of maize grains and 1.0 to 1.2 t/ha of soybean. Even productivity values of 3.0 to 4.0 t/ha of maize have been obtained in good rainfall years.

Maize-wheat and maize-mustard double crop systems have shown great promise in the Hoshiarpur, Dehra Dun, and Udaipur regions, with a production potential of 2.5 t/ha of maize, and 2.0 t/ha of wheat in Dehra Dun, and 1.0 to 1.2 t/ha of mustard in Udaipur.

Paddy-chick-pea, paddy-finger millet and paddy-wheat double crop systems are finding favour with the dryland farmers of Dehra Dun, Varanasi, Rewa, Bhubaneswar, and Ranchi.

The paddy-wheat double crop system has given, on average, 2.5 t/ha of paddy and 2.0 t/ha of wheat grains at Dehra Dun.

Pulse-based Cropping System

In areas like Anantapur and Akola, pigeon-pea forms the principal component of the intercrop system, with groundnut as the companion crop. In the Bangalore and Bhubaneswar regions, finger millet (transplanted) forms a good companion crop to the pigeon-pea, the yield of finger millet being obtained as a bonus.

Among double crop systems, green gram-safflower, cow-pea-finger millet/maize, black gram-rabi sorghum systems have shown great promise, cow-pea as the principal crop being quite promising in the Bangalore region.

Table: *Pulse-based Cropping Systems for Drylands and Their Production Potential in a Normal Season*

Cropping system	Region	Principal/First crop yield (t/ha)		Companion/Second crop yield (t/ha)	
		Grain	Edible portion	Grain	Edible portion
Intercrop system					
Pigeon-pea + groundnut	Anantapur, end Akola	1.2- 1.5	1.0- 1.2	1.0 1.2	
Pigeon-pea + finger millet (transplanted)	Bangalore, and Bhubaneswar	1.0 1.5	0.8- 1.2	0.8 1.0	0.8 1.0
Pigeon-pea (sole)		1.5	1.2		
Double crop system					
Green gram - safflower	Bijapur, Solapur, Indore, Udaipur, Ranchi, and Hissar	0.6- 0.8	0.5 0.7	1.0 1.5	
Cow-pea - finger millet	Bangalore	0.6- 0.8	0.6- 0.8	2.0 2.4	2.0 2.4
Cow-pea - maize	Bangalore	0.6- 0.8	0.6- 0.8	2.2- 2.5	2.2 2.5
Black gram - rabi sorghum	Bijapur, Solapur, Akola, and Indore	0.5- 0.6	0.4- 0.5	1.2- 1.5	1.2 1.5

Oilseed-based Cropping Systems

Red gram and castor go very well with groundnuts as the principal crop in the intercrop systems, the former (groundnut + red gram) having shown promise in the Anantapur, Solapur, Akola, Ranchi, and Rewa regions. In the Rajkot, Ranchi, and Bangalore regions, groundnut + castor system is becoming popular.

Table: *Oilseed-based Intercropping Systems for Drylands and Their Production Potential in a Normal Season*

Intercrop system	*Region*	*Yield (t/ha) of groundnut*	*Yield (t/ha) of the companion crop*
Groundnut + red gram	Anantapur, Solapur, Akola, Ranchi, and Rewa	0.9-	0 5
		1.0	0.6
Groundnut + castor	Rajkot, Ranchi, and Bangalore	2 0-	0.6
		2.5	0.8
Groundnut (sole)		2.5	

Water Potential

Water potential is the potential energy of water per unit volume relative to pure water in reference conditions. Water potential quantifies the tendency of water to move from one area to another due to osmosis, gravity, mechanical pressure, or matrix effects such as capillary action (which is caused by surface tension). Water potential has proved especially useful in understanding water movement within plants, animals, and soil. Water potential is typically expressed in potential energy per unit volume and very often is represented by the Greek letter ψ.

Water potential integrates a variety of different potential drivers of water movement, which may operate in the same or different directions. Within complex biological systems, it is common for many potential factors to be important. For example, the addition of solutes to water lowers the water's potential (makes it more negative), just as the increase in pressure increases its potential (makes it more positive). If possible, water will move from an area of higher water potential to an area that has a lower water potential. One very common example is water that contains a dissolved salt, like sea water or the solution within living cells. These solutions typically have negative water potentials, relative to the pure water reference. If there is no restriction on flow, water molecules will proceed from the locus of pure water to the more negative water potential of the solution; flow proceeds until the difference in solute potential is balanced by another force, for example, pressure potential.

Components of Water Potential

Many different factors may affect the total water potential, and the sum of these potentials determines the overall water potential and the direction of water flow:

$$\Psi = \Psi_0 + \Psi_\pi + \Psi_p + \Psi_s + \Psi_v + \Psi_m$$

where:

- Ψ_0 is the reference correction,
- Ψ_π is the solute potential,
- Ψ_p is the pressure component,
- Ψ_s is the gravimetric component,
- Ψ_v is the potential due to humidity, and
- Ψ_m is the potential due to matrix effects (e.g., fluid cohesion and surface tension.)

All of these factors are quantified as potential energies per unit volume, and different subsets of these terms may be used for particular applications (e.g., plants or soils). Different conditions are also defined as reference depending on the application: for example, in soils, the reference condition is typically defined as pure water at the soil surface.

Pressure Potential

Pressure potential is based on mechanical pressure, and is an important component of the total water potential within plant cells. Pressure potential increases as water enters a cell. As water passes through the cell wall and cell membrane, it increases the total amount of water present inside the cell, which exerts an outward pressure that is retained by the structural rigidity of the cell wall. By creating this pressure, the plant can maintain turgor, which allows the plant to keep its rigidity. Without turgor, plants lose structure and wilt.

The pressure potential in a living plant cell is usually positive. In plasmolysed cells, pressure potential is almost zero. Negative pressure potentials occur when water is pulled through an open system such as a plant xylem vessel. Withstanding negative pressure potentials (frequently called *tension*) is an important adaptation of xylem .

Osmotic Potential

Pure water is usually defined as having an osmotic potential (Ψ_π) of zero, and in this case, solute potential can never be positive. The relationship of solute concentration (in molarity) to solute potential is given by the van 't Hoff equation:

$$\Psi_\pi = -MiRT$$

where M is the concentration in molarity of the solute, is the van 't Hoff factor, the ratio of amount of particles in solution to amount of formula units dissolved, R is the ideal gas constant, and T is the absolute temperature.

For example, when a solute is dissolved in water, water molecules are less likely to diffuse away via osmosis than when there is no solute. A solution will have a lower and hence more negative water potential than that of pure water. Furthermore, the more solute molecules present, the more negative the solute potential is.

Osmotic potential has important implication for many living organisms. If a living cell with a smaller solute concentration is surrounded by a more concentrated solution, the cell will tend to lose water to the more negative water potential (Ψ_w) of the surrounding environment. This is often the case for marine organisms living in sea water and halophytic plants growing in saline environments. In the case of a plant cell, the flow of water out of the cell may eventually cause the plasma membrane to pull away from the cell wall, leading to plasmolysis. It can be measured in plant cells using the Pressure bomb. Most plants, however, have the ability to increase solute inside the cell to drive the flow of water into the cell and maintain turgor.

This effect can be used to power an osmotic power plant.

Matrix Potential (Matric Potential)

When water is in contact with solid particles (e.g., clay or sand particles within soil), adhesive intermolecular forces between the water and the solid can be large and important. The forces between the water molecules and the solid particles in combination with attraction among water molecules promote surface tension and the formation of menisci within the solid matrix. Force is then required to break these menisci. The magnitude of matrix potential depends on the distances between solid particles—the width of the menisci and the chemical composition of the solid matrix. In many cases, matrix potential can be quite large and comparable to the other components of water potential discussed above.

It is worth noting that matrix potentials are very important for plant water relations. Strong (very negative) matrix potentials bind water to soil particles within very dry soils. Plants then create even more negative matrix potentials within tiny pores in the cell walls of their leaves to extract water from the soil and allow physiological activity to continue through dry periods. Germinating seeds have a very negative matric potential. This causes water uptake in even somewhat dry soils and hydrates the dry seed. In terms of negative water potential, creosote bushes can tolerate extreme drought stress by operating fully at -50 bars of water potential and have been found living down to -120 bars.

6

Adaptations of Plants to Arid Environments

Environmental stresses of low and unpredictable precipitation, low relative humidity with desiccating winds, and high summer temperatures characterize climates of deserts and, coupled with low nutrient availability, produce severe limitations of plant growth. Despite such stresses, desert scrub communities often contain surprisingly large amounts of plant biomass, and possess remarkable diversity of plant growth forms.

The life form of a plant – whether annual, perennial, herbaceous, woody, or succulent – and the characteristics or its roots, stems, and leaves are presumed to be adaptations to the special conditions within a desert. Diversity of life forms may be considerable (Sonoran Desert) or very low (Atacama Desert).

Life forms may be classified into 4 major categories that represent strategies of adaptation, according to Solbrig and Orians, 1977.

Strategies

1. Drought-escaping plants – annuals which germinate and grow only when there is sufficient moisture available to complete their life cycle. Only their seeds persist during times of drought. Annuals.
2. Drought-evading plants – non-succulent perennials which restrict their growth activity to periods when moisture is available. Typically, they are drought-deciduous shrubs which go dormant or die back during dry periods.

3. Drought-enduring plants – evergreen shrubs. Extensive root systems coupled with various morphological and physiological adaptations of their aerial parts enable these hardy xerophytes to maintain growth even in times of extreme water stress. Creosote bush (*Larrea tridentata*)
4. Drought-resisting plants – succulent perennials. The water stored in their swollen leaves and stems is usually used very sparingly. Cacti

The major adaptation of both drought-escaping and drought-evading plants is an ability of accurately predict the wet season and to restrict their major growth and reproductive activities to the wet part of the year.

Drought-escaping Plants: Annuals

1. The proportion of annual species in desert floras is inversely related to the amount and reliability of precipitation in a region.
2. Life cycles of these small, shallow-rooted plants commence when there is water available.
   a. Went noted that no seedlings of any species germinated in the Sonoran Desert following a 10 mm rainfall; extensive germination occurred only after a rainfall of 25 mm.
   b. This was attributed to removal of inhibitors from the seed coats.
3. Winter and summer annuals are distinguished on the basis of optimum temperatures for seed germination. This varies from 15-18 C for winter annuals, and 25-30 C for summer annuals. These refer to temperature conditions prevailing at the time of precipitation.
4. Continued survival of annuals requires that adequate seed reserves be maintained through dry periods until conditions are once again suitable for germination.
5. Few seeds rot in the desert environment, but many are lost to seed predators, particularly rodents. Kangaroo rats (*Dipodomys merriami*) consume as much as 95% of the seed of Erodium cicutarium, an annual of the Mojave Desert.
   a. The highest seed densities are associated with wind shadows where sees are protected from further movement by the wind; lowest values were found in open sites and dry washes.

6. Seeds of most desert annuals have temperature or moisture controlled dormancy which may prevent germination, but seed viability is initially high. Seeds may remain viable for up to 10 years under artificial conditions.
7. Mechanisms which prevent seeds from germinating all at once can increase the chance of survival of annual species. Seed heteroblasty, in which germination requirements differ for seeds produced by the same plant, has been described for some annuals in the Saharo-Arabian Deserts.
8. Since they are not restricted by water – they grow only when water is relatively abundant – they can exploit the favourable light and temperature conditions of the desert. They require none of the morphological adaptations that other strategists require.
9. Annuals' "goal" is to grow fast (large leaves, maximal photosynthesis, heavy transpiration), flower, set seed, disperse seed, and persist through the dry period as a seed.

Drought-evading Plants

1. Drought-deciduous plants; they drop their leaves, and sometimes their stems during periods of drought.
2. Leaf production from mid-Feb. to April, and a leaf canopy persists until drought conditions become extreme in summer.
3. Leaves may be polymorphic.
    a. Leaves developed in early spring when water is available and temperatures moderate, are large and green. High photosynthetic rates.
    b. With the onset of dry period, subsequent leaves produced are smaller, and often with a pubescent covering. Lower photosynthetic rates, but lower heat load and transpiration rates.
    c. Examples: Brittlebush *(Encelia farinosa)* white bursage (*Ambrosia dumosa*).
4. Stem photosynthesis
    a. Stems lose less water than leaves (reduced transpiration).
    b. During period of low water stress, leaves are more productive than stems, but under periods of water stress, the superiority of leaves is lost.

c. Example – cheeseweed (*Hymenoclea salsola*).

d. A number of plants are essentially aphyllous (without leaves), relying on stem photosynthesis: Mormon tea (*Ephedra spp.*), spiny menodora (*Menodora spinescens*), turpentine bush (*Thamnosma montana*).

Drought-enduring Plants

1. True xerophytes, maintaining a canopy and positive net carbon gain (photosynthesis) throughout the year.
2. Example: creosote bush (*Larrea tridentata*)
   a. Compound leaves (2 leaflets) commonly live for 8-14 months, but a canopy of photosynthetically active leaves is present throughout the year.
   b. Leaves of creosote are oriented more or less vertically, parallel to the sun.
   c. Glandular trichomes secrete a resin that covers the leaf surface. The resin limits photosynthesis, but also drastically reduces transpiration.
3. Phreatophytes: These plants usually have extensive root systems which either spread through the surface soils or penetrate several metres below the surface Example – mesquite (*Prosopis*).
   a. Branching usually occurs in the capillary fringe above the water table in deep-rooted plants (phreatophytes) that are able to tap permanent ground water; while short-lived “rain roots” develop on woody surface roots in response to soil moistening.

Drought-resisting Plants – Succulents

1. Succulence is the most obvious characteristic of drought-resisting plants.
2. Desert succulents are generally shallow-rooted, allowing them to respond quickly to light rainfalls.
3. Stems are often heavily waxed to reduced cuticular water loss.
4. Leaves are often reduced to spines, and this increases the volume to surface ratio. In barrel cactus the ratio is 2.5, compared with 0.92 for a succulent leaf of an agave, and 0.01 for many non-succulent leaves. The maximum ratio is achieved by the spherical form of many cacti.

5. Spines also help to reduce heat load, and dissipate heat. Tissue temperatures below spines of the cholla cactus (*Opuntia bigelovii*) can be reduced by as much as 11 C.
6. Desert succulent are rarely killed by high temperatures, and several species of cacti and agave can withstand temperatures over 60C (140F) for short periods. However, their seedlings are especially sensitive to high-temperature injury, and establishment is often prevented in open areas where soil temperatures can rise to 80 C (176F).
7. Seedlings of saguaro and other cacti require the shade of a nurse plant, like palo verde, to survive.
8. CAM plants – stomata of succulents open in the cool of the night. Transpirational water loss is reduced, restricting CO_2 uptake to the dark hours. Assimilation products are stored at night, before being converted to photosynthates during the daylight period.
9. Water loss is very limited in desert succulents, but photosynthetic rates and, consequently, growth rates are low also. Annual growth rates in barrel cactus (*Ferocactus acanthodes*) is less than 2 cm.

Other Adaptations

1. Microphylly – typical leaves of warm desert plants are small and narrow, a design that enables leaf temperature to be near ambient temperature even when stomata are closed, so leaves can avoid lethal summertime temperatures during summer drought.
    a. Broad leaves are characteristic of plants in mesic environments or, if in the desert, in microclimatically mesic environments, like in a riparian zone, by a wash, or perched on top of a shallow water table, like the California palm (*Washingtonia filifera*).
2. Solar angles – Steeper leaf angles (parallel to solar radiation) reduces solar interception at midday during summer months, when air temperature and water loss may be most severe.
3. Solar tracking
4. Leaf reflectance / leaf absorptance
    a. Leaf absorptance in mesic communities is generally 85%.

b. Leaf absorptance in desert communities range from 60-85%, but are as low as 29% in brittle bush (*Encelia farinosa*).

c. Trichomes reduce heat load, reduce leaf temperature, reduce transpiration rates, reduces photosynthetically active radiation (negative effect), by absorbing and reflecting infra-red radiation.

5. Leaf type – Finely divided compound leaves, like those of Acacia, decrease the rate of water loss during periods of water stress, and reduce head load when stomata are closed.

What is Drought Resistance?

It is the ability of a plant to maintain favourable water balance and turgidity even exposed to drought conditions there by avoiding stress and its consequences. Stress avoidance due to morphological anatomical characteristics which themselves are the consequences of the physiological processes induced by drought these zerophytic characteristics are quantitative and vary according to environmental conditions.

A favourable water balance under drought conditions can be achieved by transpiration before as soon as stress is experienced. These are called "water savers" or.

Accelerating water uptake sufficiently so as to replenish the lost water called as "water spenders"

The mechanism for conserving water:-

1. ***Stomatal mechanism:*** Stomata of different species vary widely in their normal behaviour and range. In some species stomata remain open continuously or remain closed continuously. Many cereals open their stomata only during a short time in the early morning and remain closed during rest of the day. There is a difference in this respect between varieties of the same crop as shown by the example in two varieties of oat one is more resistant to drought open its stomata more rapidly in the early morning when moisture stress is at its minimum and photosynthesis can precede with the least loss of water (stocker 1960).

However mechanism of conserving water based on the closure of stomata will inevitable load to reduce photosynthesis and may lead to drought induced starvation injury (Leavitt, 1972).

2. ***Increased / Photosynthetic efficiency:*** On possibility for overcoming limitations on photosynthesis, imposed bicoastal closure as means for increasing resistance to loss of water by transpiration there by transpiration there by accumulations of CO2 would be at higher rate for a given stomatal opening (Hatch & stack, 1970). A number of imperfect crop plants (maize, sugarcane sorghum prose, fox tail & finger millets) (Hatch et. al. 1987) as well as certain forage species Bermuda grass (Cynodon dactyl on) Sudan grass Bahia grass (Paspalum notatum) Rhodes grass (chloris Guyana) (Murata lyama 1963) and certain A triplex sp. fixed most of CO2 into the C4 of molic and aspartic acids so called C4 dicarboxylic acid (C4) pathway.
3. ***Low rate of cuticular transpiration:*** The typical example is the cacturs. Thick cuticle results in low rate of transpiration.
4. Decreasing transpiration by a deposit of lipids layers on the surface of the leaves on exposure to moderate drought e.g. soybean (Levitt 1972).
5. ***Reduce leaf area:*** The principal means of reducing water loss of xenomorphic plants is their ability to reduce their transpiring surface. Apart from the common means of keeping the aerial parts small perhaps the simplest form of this reduction of the transpiring surface is the sealing or of leaves at the time of water stress a characteristic phenomenon exhibited by many grasses. The rolling of leaves has been shown to reduce transpiration by almost 55 percent in semi conditions and by 75 percent in desert xerophytes (Stalfect - 1956).
6. ***Leaf surface:*** Various morphological characteristics of leaves he reduce the transpiration rate and may affect survival of plants drought conditions. Leaves with thick cuticle waxy surface and the presence of spines etc. are common and effective.
7. ***Stomatal frequency and location:*** A smaller number of stomata retard the development of water deficits. In certain species, the stom are located in depression or cavity in the leaves which is feature can further reduce transpiration by limiting the impingement of currents.
8. ***Effect of awns:*** Awned varieties of wheat predominate in the drier at warmer regions and have been found to yield better than awnless one especially under drought conditions though

there are exceptions (Gurandhacher 1963). Awns have chloroplasts stomata and so as photosynthesized. It has been found that the contribution of the away to the total dry weight matter of the kernels was 12% of that the entire plant.

To Improving water uptake (MC - Donough & Gauch 1959) :-

Efficient Root System: The root systems of drought resistant plants are characterized by wide variety of apparent adaptations. These responded to such predominant soil conditions as the duration of soil dryness and the depth that is normally wet. Plants become adapted to dry conditions mainly by developing an extensive root system rather that structural modification of the roots (shields - 1958). The conceres "extensive root system" includes additional growth of secondary hair roots.

High Root to Top Ratio (R/T): A high root to top ratio is very effective mean to adoption of plants to dry conditions of the growth rate of the roots considerably exceeds that of the shoots. The transpiring surface is there by reduced while root system of the individual plant obtains it's water from a large volume of soil (Simonis 1992) has shown that an increased root top ratio may actually result in greater amount of total dry matter of plants grown under dry conditions as compared a similar ones grown with full moisture.

Difference in Osmotic Potential of Plants : Levitt (1958) has calculated a difference of 0.5% in soil moisture content that includes per manual wilting could supply a plant with enough water to keep it alive for 6 days. This could mean in certain cases the difference between survival and death.

Conservation of Water Spenders to Water Stress: Because of increased water absorption water spenders are characterized by very high rate transpiration. However as soon as the absorption rate becomes insufficient to keep up with water loss the water spenders generally develop some of the characteristics of the water savers (Cevitts - 1972).

Mitigating Stress

Mitigating Stress: Adoptions a drought basis mitigating effects of stress permit the plant to maintain a high internal water potential inspite of drought conditions. They therefore able to maintain cell tartar and growth avoid direct or indirect metabolic injury due to dehydration (Levit 1972).

Drought Tolerance

When plant is actually submitted to low water potential it can show drought tolerance by either mitigating the actual stress induced by the moisture deficiencies or by showing high degree of tolerance to stresses.

High Degree Tolerance; Resistance to Dehydration: The simplest method of avoiding drought induced damage is by resisting dehydration, preferably tot he extent .of maintaining turgur and at least by avoiding cell collapse after loss of turgur (Levit 1972) retain their turgur and therefore can continue to grow when exposed to drought stress. When plants are grown in their natural environment their osmotic potentials tend to be characteristic for each ecological group.

Agro Climatic Zones of India In General

Introductions: - The important rational planning for effective land use to promote efficient is well recognized. The ever increasing need for food to support growing population @2.1% (1860 millions) in the country demand a systematic appraisal of our soil and climatic resources to recast effective land use plan. Since the soils and climatic conditions of a region largely determine the cropping pattern and crop yields. Reliable information on agro ecological regions homogeneity in soil site conditions is the basic to maximize agricultural production on sustainable basis. This kind of systematic approach may help the country in planning and optimizing land use and preserving soils, environment.

India exhibits a variety of land scopes and climatic conditions those are reflected in the evolution of different soils and vegetation. These also exists a significant relationship among the soils, land form climate and vegetation. The object of present study is to delianate such regions as uniform as possible introspect of physiographic, climate, length of growing period (LPG) and soils for macro level and land use planning and effective transfer of agro - technology.

Agro Climatic Zones: - Agro climatic zone is a land unit in Irens of mator climate and growing period which is climatmenally suitable for a certain image of crops and cultivars (FAO 1983). An ecological region is characterized by district ecological responses to macro - climatic as expressed in vegetation and reflected fauna and equatic systems. Therefore an agro-ecological region is the land unit on the earth surface covered out of agro - climatic region, which it is super imposed on land form and the kinds of soils and soil conditions those act as modifiers of climate and LGP (Length of growing period).

With in a broad agro climatic region local conditions may result in several agro - ecosystems, each with it's own environmental conditions. However, similar agro ecosystems may develop on comparable soil, and landscape positions. Thus a small variation in climate may not result in different ecosystems, but a pronounced difference is seen when expressed in vegetation and reflected in soils.

India has been divided into 24 agro - climatic zone by Krishnan and Mukhtar Sing, in 1972 by using "Thornthwait indices".

The planning commission, as a result of mid. term appairasal of planning targets of VII plan (1985 - 90) divided the country into 15 broad agro - climatic zones based on physiographic and climate. The emphasis was given on the development of resources and their optimum utilization in a suitable manner with in the frame work of resource constraints and potentials of each region. (Khanna 1989).

Agro climatic zones of India :- (Planning Commission 1989)

1	Western Himalayan Region	Ladakh, Kashmir, Punjab, Jammu etc.brown soils & silty loam, steep slopes.
2	Eastern Himalayan Region	Arunachal Pradesh, Sikkim and Darjeeling. Manipur etc. High rainfall and high forest covers heavy soil erosion, Floods.
3	Lower Gangatic plants Regions	West Bengal Soils mostly alluvial & are prone to floods.
4	Middle Gangatic plans Region	Bihar, Uttar Pradesh, High rainfall 39% irrigation, cropping intensity 142%
5	Upper Gangatic Plains Region	North region of U.P. (32 dists) irrigated by canal & tube wells good ground water
6	Trans Gangatic plains Region	Punjab Haryana Union territory of Delhi, Highest sown area irrigated high
7	Eastern Plateaus & Hills Region	Chota Nagpur, Garhjat hills, M.P, W. Banghelkhand plateau, Orissa, soils Shallow to medium sloppy, undulating Irrigation tank & tube wells.

8	Central Plateau & hills Region	M. Pradesh
9	Western Plateau & hills Region	Sahyadry, M.S. M.P. Rainfall 904 mm Sown area 65% forest 11% irrigation 12.4%
10	Southern Plateau & Hills Region	T. Nadu, Andhra Pradesh, Karnataka, Typically semi and zone, Dry land Farming 81% Cropping Intensity 11%
11	East coast plains & hills Region	Tamil Nadu, Andhra Pradesh Orissa, Soils, alluvial, coastal sand, Irrigation
12	West coast plains & Hills Region	Sourashtra, Maharashtra, Goa, Karnataka, T. Nadu, Variety of cropping Pattern, rainfall & soil types.
13	Gujarat plains & Hills Region	Gujarat (19 dists) Low rainfall arid zone. Irrigation 32% well and tube wells.
14	Western Dry Region	Rajasthan (9 dists) Hot. Sandy desert rainfall erratic, high evaporation. Scanty vegetation, femine draughts.
15	The Island Region	Eastern Andaman, Nikobar, Western Laksh dweep. Typical equatorial, rainfall 3000 mm (9 months) forest zone undulating.

Rainfall, Its Distribution and Its Effectiveness In Rainfed Agriculture

Distribution of Rainfall: - The amount of rainfall received at periodic intervals like weeks, months, seasons etc. indicate distribution. In addition distribution of rainfall can be known by the length of dry spell, wet spells land rainy days. Distribution of rainfall is more important than total rainfall.

Rainfall pattern at three locations:-

Sr. No.	*Index*	*Hyderabad*	*Solapur*	*Dhule*
1	Annual rainfall (mm)	764	742	625
2	Seasonal rainfall (mm)	580	556	450
3	Coefficient of variation%			
4	PE (mm)	1757	1802	1502
5	Growing season	130	148	125
6	Soil	Vertisols	Vertisols	Medium deep

From the above data the growing season is slightly less at Hyderabad as compared to Solapur, However rainy season crops are more successful at Hyderabad and annual yields range 50 to 70 q/ha. While at Solapur and Dhule rainy season crops are risky and annual yields range from 10 to 12 q/ha. Low yields at Solapur and Dhule are mainly due to discontinuous rains or long breaks in rainfall during crop growth period.

Rainfall distribution is based on:

1. Weekly or monthly rainfall will give distribution of rainfall in weeks during a crop season.
2. Wet and dry spells - A wet spell is a number of continuous days of rainfall. A dry spell is a number od continuous rainless days.
3. Rainy days: If the rainfall received is more than 2.5 mm on any day. This particular day is called rainy day.
4. Periodicity of rainfall.
5. Onset of monsoon.
6. Recurrence of rainfall events.
7. Dependability of rainfall
8. Certificient of variation. If C.V. is more variation in rainfall is more and vice - a - versa.
9. Length of the growing season (LGS): If LGS is less a short duration crop should be selected. L.G.S. depends on duration of rainy season and moisture retention.

Rainfall being a single most important factor for success of crops in the dry farming areas. It is generally known that India receives its annual rainfall by the particular phenomenon called monsoon which consists of series of cyclones those arise in the Indian Ocean. These travel in the North East direction and enter the peninsular India along the Western coast. These cyclones occur from June to Sept. is known as south West monsoon. This is followed by second third and fourth rainy season during periods from Oct. to Nov., Dec. To Feb. and March to May respectively. South West Monsoon is the most important as it covers major parts of India and brings bulk of the total annual rainfall.

The North East of Returning monsoon: By the end of Sept. South West Monsoon ceases to penetrate North West India but continues a full month longer in Bengal. On account of south East North easterly winds being to flow on the Eastern coast. Some times some of these cyclones penetrate In land and give supplementary rainfall to dry region of the plateau of the peninsular India. This is known as returning monsoon.

Precipitation and Its Factors

Precipitation is reaching of atmospheric humidity either as rain or snow to the ground. OR Precipitation can be defined as earth word falling of water drops of ice particles that have formed by rapid condensation in the atmosphere and are too large to remain suspended in the atmosphere.

Factors influencing precipitation:-

1. Only blowing of winds coming even over the sea is not enough to produce precipitation.
2. Horizontal movement is not conductive to precipitation.
3. The rain bearing clouds, hills, mountains, slanting slopes of the river ralleys lake dynamic cooling of the clouds.
4. Water vapour in atmosphere and moat conditions which promote greater precipitation.
5. Regiour covered with thick forest contributes more water vapour by transportation and thus provide favourable condition for preciption.
6. The prevalence of dry winds, higher temperature absence of barriers and cutting of monsoon currents these are unfavourable for precipitation.
7. Long & short breaks in the monsoon caused due to prevalence of dry winds slowing over land or desert plains from North East.
8. On the other hand geographical position, physical configuration and meteorological conditions are responsible for precipitation.

Type of Rainfall In Dry Areas

1. First type rainfall: Rainfall receives from south west Monsoon. Rainfall receives up to 60% in the first three months viz. June - July - August - Rontak Jodhpur Jalgaon.
2. Second type rainfall: Rainfall receives from south - West Monsoon (40 to 55%) and supplemented with North East Monsoon (40 to 50%) Pune. Wal A Nagar Raichur - Maximum rainfall receives in July & Sept.

3. Third type rainfall: Rainfall receives from North East Monsoon (60%) Solapur Bijapur a Karnataka.
4. Four type rainfall: Rainfall receives uniformly (Well distributed) from Both Monsoon currents, places of rainfall - Chennai. Total rainfall received in 5 to 6 months.

Decennial rainfall: The mean total rainfall received during past 10 years:

Winds coming over land surface from the North - East are dry and cold and man cause of breaks in the monsoon or they tend to decrease the rainful of a tract by diluting the moisture laden masses of the atmosphere. Indian be divided into three zones of the basis of rainfall.

A) Heavy rainfall zone: above 1250 mm.

B) Moa crate rainfall zone: 750 to 1250 mm.

C) We rainfall zone: Less than 750 mm annual rainfall.

The average rainfall of Solapur varies from 500 to 720 mm and had bimodal stribution. The first peak is usually experienced during June and Second during Sept. Rainfall during Sept. is more assured and is in the rage of 150 to 200 mm. Even though Monsoon sets in by the end of June, July and August are characterized by dry spells of varying duration (2 to 8 weeks at stretch) and frequencies 1 to 5. Usually dry speels of more that 4 weeks duration or 3 dry spells of 2 week duration result in failure off Kharif crops. Such occasions are observed twice in fire years. Usually high wind velocity (18 to 20 km / hr)

At Solapur under dry land areas year to year fluctuatins are so much that there is no guarantee of a fixed quantity of rainfall. Generally rainfall starts in late June to early July. There is depression during late July to early August. Again there is good amount of rainfall in last Aug. and Sept. The rainfall totally recedes by mid October. This is the usual pattern of rainfall in draught prone areas. The probability of rainfall is more than half the normal is fairly good. (P = 0.58) during September.

Techniques of Soil and Water Conservation In Rainfed Agriculture

Soil and water are most essential for the growth and sustenance of plant life. Soil is important as it provides, foothold for plants and majority of nutrients needed by them. Water is essential as it forms larger part of the living matter and acts as a nutrient carrier.

Though, both soil and water a sources are available in plenty, they are not distributed equally in quality and quantity in every part of the world and are not inexhaustible. Their abuse would mean a great loss resulting in poverty. It takes centuries to form one inch layer of soil, but it does not take long to lose it by erosion. Research work carried out in Maharashtra State which has an undulating topography, has shown that loss of soil from unprotected land is as much as 125 tons per hectare every year and may be as high as 300 tons in a single year. The weight of one hectare of soil 2.5 cm. deep is about 325 tons.

Similarly, rain water which can sustain a good crop, if not conserved properly will not only cause scarcity and famine, but also wash way the soil which is a valuable national asset. There are many examples which show how once fertile plains and valleys have become deserts or barren lands due to neglect by mankind. It is, therefore, the prime responsibility of each generation to conserve soil which is the main capital of the farmer as well as the nation, at all costs and pass it on in good condition from one generation to another, so that the posterity will not blame them.

Soil and water conservation cannot be achieved only by individual efforts. The problem is too big, involving collective efforts on the part of farmers, technicians and Government. Recognizing the seriousness of erosion problem, the Central Government established the Central Board of Soil Conservation to assist the States and River valley Projects. It has established soil conservation research stations at Dehradun, Kotah, Ootacamand, Bellary, Vasad and Jodhpur, arranges for training of technical personnel and also served as clearing house for soil conservation information.

On account of chronic scarcity conditions prevailing over 3 major portions of the Deccan tract. Soil and water conservation research was started in 1924 and soil conservation work was taken up on a large scale in Maharashtra from 1943 - 44 onwards. At present Maharashtra contributes nearly 50 p.c., of the total progress in respect of soil conservation measures in the country.

Water Conservation

Along with soil water is another important factor essential for all life and production of food. The main source of water is precipitation. In India precipitation is not property distributed throughout the year. It is received within a few months of rainy season and that too in a critic manner. It may rain 50 to 125 mm. in one day causing flood and damage to crops and then there may be a dry spell for some days when crops may begin to will. Hence proper conservation of water as well as collecting surplus water in tanks and reservoirs of letting it out into the rivers assume great importance. To understand water conservation it is necessary to study the hydrologic cycle.

Hydrologic Cycle

Water moves in a continuous cycle from ocean to clouds to earth and back to ocean. The water in the ocean is converted into vapour by the heat of the sun and this vapour moves in the form of clouds over the land and condenses into rain. Some rainwater enters the soil while the rest flows into streams and rivers and is either stored in tanks and reservoirs or allowed to go back to sea. Of the water that enters the soil some is stored for use by the plant some gets evaporated from the surface of the soil and some moves down to replenish the water table. This becomes the source of water for wells and springs. This cyclic movement of water is known as the hydrologic cycle. The water that is held by the soil in available form is essential for plant growth. It is not yet possible for the man to control rainfall but its infiltration and run off can be regulated to a great extent by improved management practices. Efforts should be made to store rain water either in the soil or in the reservoirs when it is in plenty and carry over to periods when rain water is not available. The former is known as conservation of water in soil, while the latter, conservation in reservoirs.

Loss of Water From Soil

Water is lost from soil in four ways:

1. Surface run off.
2. Downward movement of drainage.
3. Evaporation from soil surface.
4. Transpiration through leaves of plants.

Out of this loss of water through run off is the largest and is also the most damaging as it causes erosion main factor in conserving moisture relates to increasing infiltration and storage capacity of soil and reducing run off and evaporation. Uncontrolled water is the main cause of soil erosion. Almost all methods that deal with soil conservation are in principle the methods to control and conserve water. Soil and water conservation are, therefore, dealt together, Sufficient studies on all phases of hydrologic cycle such as evaporation, precipitation, run off, infiltration and deep percolation which occur simultaneously have not been carried out in India as yet, though 2 beginning has been made at many agricultural research stations.

Soil and Water Conservation Methods

The loss of soil and water under natural vegetation is the lowest. But lands must be cultivated and grown with crops to produce food. This can be done without much harm to the soil if proper soil and water conservation methods are followed. Such methods aim at encouraging water to infiltrate into the soil, reduce its velocity and check run off losses.

The most common soil and water conservation methods are

A) Management practice viz.

   a) Strip cropping,

   b) Mulching,

   c) Crop rotation,

   d) Contour cultivation,

   e) Planting of grasses for stabilizing bunds,

   f) Planting of trees and a forestation,

   g) Cashew nut plantation, and

B) Mechanical practices such as

   a) Bunding,

   b) Terracing,

   c) Gully or nala control,

   d) Control of stream and river banks.

Soil and Water Conservation Methods - Management Practices

Strip Cropping: This consists of growing erosion permitting crops and erosion resisting crops in alternate strips. The erosion permitting crops are cotton, jawar, bajara, etc. which allow the run off water to flow freely within the rows. The erosions resisting crops are mostly legumes like groundnut, much (Phaseolus aconitiolius), hulgn (Dolichos biflorus), Sow bear (Glycine max) which spread and cover the soil and do not allow run off water to carry much soil with it the soil which flows from the strips growing erosion permiuming crops is caught by the alternating springs.

In selecting a suitable legume crop it should be seen that the maximum canopy and root development of the crop coincide with the period of high intensity of rainfall.

Mulching: A mulch is natural or artificially applied layer of plant residues or other material on the surface of the soil with the object of moisture conservation, temperature control, prevention of surface compaction or crust formation, reduction of run off and erosion, improvement in soil structure and weed control. Artificial mulches of different kinds such as *Jowar* or *bajara* stubbles, stubbles, paddy straw or husk, sawdust etc., increase absorption of water and minimize evaporation. They also control run off and soil losses.

Rotation of Crops: Rotation means growing a set of crops in a regular succession over the same field within a specified period of time. Continuous growing of Jowar or bajara crop causes more erosion, but if followed by a legume crop viz., hulga, matki or gram which covers the soil is causes less erosion. Rotation also helps in removal of plant nutrients in a uniform way from future depth of soil and in maintaining the fertility of the soil in dry farming region of Maharashtra adoption of gram Jowar rotation not only helps in conservation of moisture but also in increasing the crop yields the beneficial effect of rotation.

Contour Cultivation: Tillage operations viz., ploughing, harrowing, sowing and Interculture should be done across the slope of land. This will help in creating obstructions to the flow of water at every furrow, which acts like a small bund and results in uniform distribution of water. This helps more initration of water less run off and erosion, and gives higher crop yield. Any cultivation done along the slope will accelerate golly formation, more run off and erosion and consequently permanent damage to land.

Planting of Grasses for Stabilizing Bunds: Grasses prevent soil erosion and improve soil structure. The entire soil mass is penetrated by countless roots and soil aggregates and particles are enmeshed by the root system. Grasses should be grown on bunds which are not suitable for cultivation, both for checking erosion and providing pasture for cattle. Several grasses as well as legumes were tried on bunds at the Agricultural Research Station; Solapur which receives about 600 mm. of rainfall to see which of them will withstand drought conditions, give maximum root growth and canopy coverage, and stabilize bunds effectively. It was observed that anjan planted with spacing of 15 x 15 cm., produced the highest quantity of roots, followed by marvel - 8, Rhodes, thin Napier, blue panic, and *kusal* .Legumes do not have many roots but produce better canopy within a short period, while grasses are under the process of establishment. Planting of legumes mixed with grasses is, therefore, advantageous in preventing soil erosion in initial stages.

Planting of Trees and Afforestation: Forests conserve soil and water quite effectively. They not only obstruct the flow of water, but the falling leaves provide organic matter which increases the water holding capacity of the soil. If tree planting is done in the planned manner in open areas, it will serve as good wind break and if done along the banks of streams and rivers, it will regulate their flow. Farm forestry is another important aspect in soil and water conservation. The danger caused by deforestation has been only recently appreciated and a big plantation programme of hybrid eucalyptus, teak, casurina has been taken up by the Forest Departments in reserve forests, catchments areas of irrigation projects and on Government waste lands Vanamahotsava is also observed every year in the early part of monsoon and millions of trees are planted by the public with the help of the maff of the Agricultural and Forest Departments local bodies like Zillah Paris had, Panchayat Samitis and Gram Panchayat What is however important is to pay proper attention planted.

Cashew Ant Plantation : In coastal districts of Maharashtra which receive more than, 1,250 mm. rainfall a new programme of cahsewnut plantation has been undertaken from 1963 - 64 on hills having slope between 10 and 20 p.c.The sea breeze is conducive to the growth of cahsewnut plants and they do not require much aftercare once they establish in the soil. Staggered trenches of 300 x 30 x 30 cm. size are dug on contours at a distance of 6m. Cashew plants are raised in polythene bagsin nurseries and two months olds a plings are planted on the lower side of the trench with plant to plant distance of 6m.

Soil and Water Conservation Methods - Mechanical Practices

The above measures control erosion by good management practices. Bunding, terracing, *gully* or *nala* control, and construction of tanks and bandharas are mechanical measures requiring engineering techniques and structures. They reduce run off and impound water for longer time to help infiltration into the soil. Their construction and design will depend upon rainfall, soil slope and such other factors. These measures are costly but if properly maintamed will improve the land over a long period of time.

Bunding

Block Bunding: Bunding for control of soil erosion and conservation of surface run off was known to farmers for centuries. It was not uncommon to find Tals i.e., big bunds across large blocks of sloping lands. These bunds are constructed of earth or stone or both, at a great cost, to impound water and arrest soil washed from the fields lying above. They are high and broad enough to withstand the force of water from the catchments. Water is let out at the end of the monsoon and land which has received fertile silt is sown with crops. Such type of big bloods bunds are not constructed now as contour bunding has been taken up on catchments basis.

Contour Bunding: It consists of construction of a series of earthen bunds of suitable sizes along contours at a lateral distance of every 60 mm or a fall of 1 to 1.5 m. The shope of land is thus broken into smaller and more level compartments which hold soil as well as rain water. It has been estimated that about 75 million hectares of land i.e. about one fourth of the common land surfaces suffering from soil erosion. In Maharashtra State, the problem is more acute and it is estimated that out of 186 lakh hectares about 144 lakh hectares require bunding. The planning Commission has, therefore laid great stress on contour bunding programme, because bunding alone has been found to increase crop yield by 20 to 30 p.c.

The size, cross- section and interbund spacing depend upon the nature of rainfall, soil and slope of the area. In order to improve the technique of bunding, studies have been carried out in respect of spacing of bunds, shrinkage of bund sections and hydraulic gradients and kind and location of outlets etc. in different soil and rainfall conditions of Maharashtra State. On the basis of such studies it has been observed that the spacing between bunds should not be allowed to exceed 1.5m. Vertical drop or 67.5 m. lateral spacing.

Graded Bunding: In high rainfall areas, while conservation of soil is important, drainage of surplus water has to be attended to, for avoiding waterlogged condition of soil. The bunds are therefore, slightly graded longitudinally about 7.5 cm. per running 33 m. to prevent safe disposal of water into the nala. The cross sections into for safe removal of excess run off water it is essential to provide suitable waste water or outlet structures at proper places so that no damage is done to bunds in case heavy precipitation is received on any single day. Normally stone outlets are provided low rainfalls are as. Channel weirsor pipe outlets may also be provided. Grass outlets have been found to be effective and cheaper in heavy soil. The crest wall should be 30 cm. above the contour level and its length should be so designed as to discharge the surplus water from the maximum intensity of rainfall with are asonable period. 1,250 mm. Terrace bunds consist of comparatively narrow embankments constructed at intervals across the shope and the vertical spacing between bunds may vary from 1to

2 m., depending upon the slope, type of soil, rainfall etc. Bench terracing is done when gradient is stceper than 10 p.c. as in hilly ranges of Himalayas, Sahyadry etc. and consists of a series of step like platforms along contours. These terraces are like table tops sloping outwards and are provided with stone wateweirs to drain away surplus water. Angular and big boulders should be used for terrace outlets because round and small boulders will slip and get dislodged under the gushing water.

Gully or *nala* control: Gully or nala control is very essential to prevent its extension and further destruction of cultivated lands and grasslands. The sloping sides are planted with grass and trees. Suitable temporary and permanent structures such as check dams, overflow dams, drop structures are also provided. Small gullies can be stabilized by converting them into paddy fields. So far 17,034 nalas have been controlled and the target for sixth plan (1980 - 85) period is 2005.

Control of stream and river banks: Vulnerable sharp bends nalas by the sides of the rods and river bends near village sites cause considerable damage to property. These should be protected by providing spurs, jetties, rivets and retaining walls. Adjoining areas should be stabilized under permanent vegetation. Spurs are constructed at an angle to reduce the velocity of water and there by enabling the flood water to flow away but deposit coarse sand which will cause obstruction to successive water currents from cutting into the bank and thus straightening their course.

Runoff, and Their Types

Definition:" It is that portion of rainfall, which makes its way towards streams, rivers etc. After satisfying the initial losses etc. is called as runoff.

Types of Runoff:

1. Surface runoff.
2. Sub - surface runoff, and
3. Base flow.

1. Surface Runoff: It is that portion of rainfall which enters the stream immediately after the rainfall. It occurs. When all losses are satisfied and if rain is still continued, with the rate greater than in filtration rate; at this stage the excess water makes a head over the ground surface (surface detention) which tends to move from one place to another, known as overland flow. As soon as the overland flow joins to the streams, channels or oceans, termed as surface runoff.

2. Sub - surface Runoff: That part of rainfall, which first leaches into the soil and moves laterally without joining the water - table to the Streams Rivers or oceans is known as sub - surface runoff. Sometimes sub - surface runoff is also aerated under service ninoff due to reason that it takes very title time to reach the river or channel in comparision to ground water. The sub - surface runoff is usually referred as interflow.
3. Base flow: It is delays flow, defined as that part of rainfall which after talling on the ground surface in fill rated into the soil and meets so the water table and flow to the streams oceans etc. The movement of water in this type of runoff is very slow that is why it is also referred as delayed runoff. It takes a long time to join the rivers or oceans. Some times base flow is also known as ground water flow.

Thus,

Total Runoff = Surface runoff + Base flow (Including sub - surface runoff)

Factors Affecting Runoff

The runoff rate and its volume from an area, mainly in influenced by following tow factors:

a) Climatic factors and
b) Physiographic factors

Factors Affecting Runoff - Climatic Factors:

The climatic factors of the watershed affecting the runoff are mainly associated with the characteristics of precipitation, which include.

1. Type of precipitation
2. Rainfall intensity
3. Forms of precipitation
4. Duration of rainfall
5. Rainfall distribution
6. Direction of prevailing wind and
7. Other climatic factors.

Type of Precipitation: Types of precipitation have a great effect on the runoff. For example a precipitation which occurs in form of rainfall, starts immediately in from of surface flow over the land surface, depending upon its intensity as well as magnitude, while a precipitation which takes place in form of snow or hails the flow of water on ground surface will not take place immediately, but after melting of the same. During the time interval of their melting the melted water infiltrates into the soil and results a very little surface runoff generation.

Rainfall Intensity: The intensity of rainfall has a dominating effect on runoff yield. If rainfall intensity is greater than infiltration rate of the soil the surface runoff takes place very shortly while in case of low intensity rainfall, where is found a reverse trend of the same. Thus high intensities rainfall yield higher runoff and vice-versa.

Duration of Rainfall: Rainfall duration is directly related to the volume of runoff due to the fact, that infiltration rate of the soil goeson decreasing with the duration of rainfall till it attains constant rate. As a result of this even a mild intensity rainfall lasting for longer duration may yield a coverderable amount of runoff.

Rainfall Distribution: Runoff them a water sheed depends very much on the distribution of rainfall. The rainfall distribution for this purpose can but expressed by a team "distribution coefficient which may be defined as the ratio of maximum rainfall at a point to the mean rainfall of the watershed. For a given total rainfall, if all other conditions are the same, the greater the value of distribution coefficient, greater will be the peak runoff and vice - versa. However, for the same distribution coefficient, the peak runoff would be resulted from the storm, falling on the lower part of the basin i.e. near the outlet.

Direction of Prevailing Wind: The direction of prevailing wind, affected greatly the runoff flow. If the direction of prevailing wind is same as the drainage system then it has great influence on the resulting peak flow and also on the duration of surface flow, to reach at the outlet. A storm moving in the direction of stream slope produces a higher peak in shorter period of time than a storm moving in opposite direction.

Other Climatic Factors: The other climatic factors, such as temperature wind velocity, relative humidity, annual rainfall etc. affect the water losses from the watershed area to a great extent and thus the runoff is also affecter accordingly. If the losses are more the runoff will be less and vice -versa.

Factors Affecting Runoff - Physiographic Factors

Physiographic Factors: Physiographic factors of watershed consist of both, the watershed as with as channel characteristics. The different characteristics of watershed and channel, which affect the runoff, are listed below.

1. Size of watershed
2. Shape of watershed
3. Slope of watershed
4. Orientation of watershed
5. Land use
6. Soil moisture
7. Soil type
8. Topographic characteristics, and
9. Drainage Density.

Size of Watershed: Regarding the size of watershed, if all other factor including depth and intensity of rainfall are being same them two watershed irrespective of their size, will produce about the same amount of runoff .However a large watershed takes longer time for raining the runoff to the outlet as result the peak flow expressed are depth is being smaller and vise versa.

Shape of Watershed:The shape of watershed has a great effect of runoff. The watershed shape is generally expressed by the terms "from factor and "compactness coefficient".

Shope of Watershed: The shope of the watershed has an important roel over runoff but its effect is complex. It controls the time of overland flow and time of concentration of rainfall in the drainage channel which provide accumulative effect on resulting peak runoff. For example in case of a sloppy watershed. The time to reach the flow at outlet is less, because of greater runoff velocity which results into formation of peak runoff very soon and vice -versa.

Orientation of Watershed: This factor affects the evaporation and transpiration losses from the area by making influence on the amount of heat to the received from the sun. The north or south orientation of watershed, affects the time of melting of collected snow. In a mountainous watershed the part of wind ward side of the mountain receives high intensity of rainfall resulting into more runoff yield while the part of watershed typing towards leeward side has reverse find of the same.

Land Use: The land use pattern and land management practices used have great effect on the runoff yield. For example an area which is under forest cover, where a thick layer of much of leaves and grasses etc. has peen accumulated there formed a little surface runoff due to the fact that more rain water is absorbed by the soil. While in a barren field where not any type of cover is available a reverse trend is obtained.

Soil Moisture: The magnitude of runoff yield depends on the amount of moisture present in the soil at the time of rainfall. If rain occurs over the soil which has more moisture the infiltration rate becomes very less which results in more runoff yield. Similarly if the rain occurs after a long dry spell of time when the soil is dry, causing to absorb huge amount of rain water. In on the other hand, if the rain occurs in a close succession as in the rainy season; runoff yield has reverse effect.

Soil Type: In the watershed surface runoff is greatly influenced by the soil type as loose of water from the soil is very much dependent on inflientation rate which varies with the types of soil.

Topographic Characteristics: Topographic characteristics include mores topographical features of watershed which create their effect on runoff it is mainly undulating nature of the reason that runoff water gets additional power to flow due to slope of the surface and altitude time to infiltrate the water into solid.

Regarding channel characteristics to describe their effect on runoff the channel cross-section, roughness storage and channel density are mainly considered. These also have significant effect on runoff.

Drainage Density: The rain age density is defined as the nation of the trial channel length in the watershed to two total watershed areas it is expressed at.

$$\text{Drainage density} = \frac{\text{(Tranned length (Total))}}{\text{Watered area}}$$

$$\text{D.D.} = \frac{\text{I}}{\text{A}}$$

A watershed having greater D.D. and incites formation of peak rain off very shortly to that of lesser D.D. watershed.

Different Agronomical Practices for Soil and Water Conservation

Conservation In Rainfed Areas

Soil conservation is a preservation technique, in which deterioration of soil and its losses are conserved by using it within its capabilities and applying conservation techniques for protection as well as improvement of soil. In hilly regions. Where land topography has steep slope and is subjected to erosion problem the vegetation cannot get established. Lack of the vegetative cover on sloppy soil surface accelerates the erosion and a large amount of soil is transported into the stream through runoff. In addition, the uncovered sloppy land also a cause extensive damage to the cultivable land at foothill through exporsition of sedimentson them.Sediment disposition covers the top fertilesoil layerand thus makes them unsuitable for cultivation.

Under this circumstance it becomes very necessary to treat such areas by adopting appropriate agronomical measures, so that they can be reclothed with negetations. The vegetation helps in reducing the surface runoff and soil cravsion both. The agronomical measures include contouring strip cropping and niluge practices to control they soil erosion. The use of these measures is entirely dependent upon the soil types land shope and rainfall characteristics.

In soil and water conservation programmes, the agronomical practices are counted as second line of defence the first being mechanical or engineering measures which are employed to arrest the soil erosion immediately. The role of agronomic measure is more economical long-lasting and effective. Always it is advisable to used but when its use is either inadequate or not sulpewant to achieve the goal of erosion control then use of mechanical measures to control erosion is recommended.

The agronomical measures are referred by the practices of growing vegetables on mild sloppy lanks to cover them and to control the erosion from there in living vegetation above the soil surface dissipates the crove power of agents either they are water or wind In case of water erosion it affects by several ways such as by enhancing infiltration rate and relucing together and thereby reducing runoff velocity to scour the soil particles screening the eroded particles to reach them into the channels or reservoirs; by dissipating the kinetic energy of falling raindrops and thus reducing the splash erosion. The effect of vegetation on wind erosion is also significant as it directly makes a hinderance in blowing path and thus deflecting the wind current at some distance away towards down stream side. The wind - strip cropping is a well known agronomical practice comployed for controlling the wind erosion in wind erosion susceptible areas.

The role of agronomical measures in achieve of soil & water conservation, has immense importance, perhaps much more than the others. It can be explained by considering the Universal Soil Loss Equation (A = R K L S C P) in which agronomical practices reflect the factor of crop management (C). The other factors such as R & K are the natural factor; we do not have any control on them. The L S and P factors may have value as I under worst conditions; although these can be reduced maximum up to 0.5 by applying an ideal soil and water conservation measures. The factor 'C' which is crop management factor has value as I for worst conditions, but it can be reduced up to 0.02. At this small value of C, the soil loss can be minimized up to one - fifteenth which is about 10.25 times more than the other factors. Looking this important effect of agronomical measures on soil loss, its scope is assumed to be more dominating in soil and water conservation programmes.

1. Contouring
2. Trip Cropping
3. Tillage Practices

These are the important agronomical practices employed for controlling the soil erosion from sloppy areas. Basically these measures create an obstruction in flow path of surface runoff by making the land surfaces rough due to channels ridges etc. formed under them. Each of these measures also have a direct relation with the infiltration rate and thereby presence of moisture in the soil profile. Infiltration rate is an effective factor in reducing the surface runoff and soil loss.

7

Different Agronomical Practices for Soil and Water Conservation

Conservation In Rainfed Areas

Soil conservation is a preservation technique, in which deterioration of soil and its losses are conserved by using it within its capabilities and applying conservation techniques for protection as well as improvement of soil. In hilly regions. Where land topography has steep slope and is subjected to erosion problem the vegetation cannot get established. Lack of the vegetative cover on sloppy soil surface accelerates the erosion and a large amount of soil is transported into the stream through runoff. In addition, the uncovered sloppy land also a cause extensive damage to the cultivable land at foothill through exporsition of sedimentson them.Sediment disposition covers the top fertilesoil layerand thus makes them unsuitable for cultivation.

Under this circumstance it becomes very necessary to treat such areas by adopting appropriate agronomical measures, so that they can be reclothed with negetations. The vegetation helps in reducing the surface runoff and soil cravsion both. The agronomical measures include contouring strip cropping and niluge practices to control they soil erosion. The use of these measures is entirely dependent upon the soil types land shope and rainfall characteristics.

In soil and water conservation programmes, the agronomical practices are counted as second line of defence the first being mechanical or engineering measures which are employed to arrest the soil erosion immediately. The role of agronomic measure is more economical long-lasting and effective. Always it is advisable to used but when its use is either inadequate or not sulpewant to achieve the goal of erosion control then use of mechanical measures to control erosion is recommended.

The agronomical measures are referred by the practices of growing vegetables on mild sloppy lanks to cover them and to control the erosion from there in living vegetation above the soil surface dissipates the crove power of agents either they are water or wind In case of water erosion it affects by several ways such as by enhancing infiltration rate and relucing together and thereby reducing runoff velocity to scour the soil particles screening the eroded particles to reach them into the channels or reservoirs; by dissipating the kinetic energy of falling raindrops and thus reducing the splash erosion. The effect of vegetation on wind erosion is also significant as it directly makes a hinderance in blowing path and thus deflecting the wind current at some distance away towards down stream side. The wind - strip cropping is a well known agronomical practice complo yed for controlling the wind erosion in wind erosion susceptible areas.

The role of agronomical measures in achieve of soil & water conservation, has immense importance, perhaps much more than the others. It can be explained by considering the Universal Soil Loss Equation (A = R K L S C P) in which agronomical practices reflect the factor of crop management (C). The other factors such as R & K are the natural factor; we do not have any control on them. The L S and P factors may have value as I under worst conditions; although these can be reduced maximum up to 0.5 by applying an ideal soil and water conservation measures. The factor 'C' which is crop management factor has value as I for worst conditions, but it can be reduced up to 0.02. At this small value of C, the soil loss can be minimized up to one - fifteenth which is about 10.25 times more than the other factors. Looking this important effect of agronomical measures on soil loss, its scope is assumed to be more dominating in soil and water conservation programmes.

1. Contouring
2. Trip Cropping
3. Tillage Practices

These are the important agronomical practices employed for controlling the soil erosion from sloppy areas. Basically these measures create an obstruction in flow path of surface runoff by making the land surfaces rough due to channels ridges etc. formed under them. Each of these measures also have a direct relation with the infiltration rate and thereby presence of moisture in the soil profile. Infiltration rate is an effective factor in reducing the surface runoff and soil loss.

Strip Cropping

Strip cropping is also a kind of agronomical practice in which ordinary crops are planted / grown in form of relatively narrow strips across the land slope. These strips are so arranged that the strip crops should always be separated by strips of close growing and erosion resistance crops.

Strip cropping used as a technique for erosion control is a most effective method in certain soils and topography. This method becomes more effective for erosion control, which it is followed with crop rotations in the area where terraces are not practically feasible due to the fact that the length of slope is divided into different small segments. The strip crops check the surface runoff and force them to infiltrate into the soil, thereby facilitates to the conservation of rain water. Strip cropping is more intensive practice for conserving the rain water than contouring (i.e. about twice as effective as contouring) but it does not involve greater effect on soil erosion as terracing and bunding. Generally the use of strip cropping practice for soil conservation is decided in those areas where length of slope is not too longer.

Strip - cropping to control soil erosion caused by runoff derives its effectiveness mainly from following two factors:

a) Reducing the runoff flowing through the close - growing sod strips.

b) Increasing the infiltration rate of the soil under cover condition.

The reduction of runoff velocity between the row strips is achieved by making an observation in the flow path. The observations created by row crotion are also responsible to dissipate the kinetic energy of flow checking the flow of surface water. From field studies it has been observed that the strip cropping on the contour plays a key role in conserving the soil and water, when combined with terracing. The width of these strips depends on the topographical features of the area.

Field Strip Cropping: It is modified form of contour strip cropping in which crop strips are laid parallel across the land slope but not always exactly on the contour may be changed. This type of strip cropping is frequently used only where the topography is either too imegular or undulating as it makes accurate layout of contour strip cropping, impractical. The depressed areas should be avoided from field strip cropping they may be left for establishing the grassed water ways.

Buffer Strip Cropping: In buffer strip cropping the strips of grasses or legume crops are laid between contour strip crops in regular rotation. The width of these strips may or may not be even. The buffer strips are usually 2 to 4 m wide and are placed at 10 to 20 m metres. They can also be placed on critical stops of the field (The main purpose of buffer strip cropping is to provide a protection to the land from soil croson.)

Wind Strip Cropping: In wind strip cropping system the strip crops of uniform width are laid at right angles to the direction of preventing winds without regard of the contour. The main objective of this system is to control the wind crosion father than water crosion. This cropping is recommended for level or nearly level topography where wind crosion is more effective.

A guideline for deciding the width of wind strip - cropping can be have from the values given in table.

Table: *Recommended strip widths for wind stripCropping (FAO 1965)*

Soill Types	***Strip width (m)***
Sandy soil	6.0
Loamy sand	7.0
Sandy loam	30.0
Loam	75.0
Silt loam	85.0
Clay loam	105.0

Layout of Contour Strip Cropping: In layout of a field for contour strip cropping the first step is to decide the width of strips at narrower points let the minimum width is assumed to be as 25 m. The next step is to establish a point for locating the contour line that will form the lower boundary of first strip. This point is located at 25 m apart from the top boundary of field by measuring along the steepest part of the stop. Now a contour line is drawn passing through this point up to the field boundary. This procedure is repeated until the entire field is laid out.

Width of Strips: It varies with the degree and length of land slope allowable soil loss soil types arrangements of crops grown in rotation and size of farm equipments used in tem aced fields the width of strip is adjusted according to the terrace interval but in untraced areas narrow width than the standard terrace interval is frequently used. In general steeper the slope narrower will be the strips of cultivated and dense growing crops both. An approximate range of trip widths based on average land slope and soil types is given in table.

Table: *Approximate range of strip width*

Sr.No.	*Percent Trand note*	*Width of strips (m)*		
	(average)	*Good soil*	*Fair soil*	*Poor soil*
1	2	51	42	33
2	5	42	33	25
3	8	33	25	17
4	11	25	17	17

A buffer strip is more or less in a permanent contour strip usually varies in width which is normally kept between 3 to 5 m.

Crop Rotation: Crop rotations can be more effective for controlling soil crosion accompanied with strip cropping system. It can be used on the same piece of land by growing tilled crops; small grain crops hay crops or grasses either under a strip cropping system or a separate field system. In areas where perennial grasses and legumes are not feasible to grow, the row crops of small grain and annual legume crops can also be grown in strips.

It is a general rule that no two cultivated strips should have the same planting or harvesting dates. The sequence of crops should be in such a manner that there could be form a dense - fibrous root system to hold the soil and retard the erosion, until the roots are croken down by tillage operations.

All these activities of crop rotation also increase the organic matter in the soil thereby the physical condition of the soil become improved ultimately soil absorbs more water and also increases the capability of soil to resist the erosion.

Under use of crop rotation practices for controlling soil crosion, the simplest way to combine different crops in roa form and grow them in consecutive rotations.

The frequency with which row crops should be grown depends upon the severity of crosion, taking place in the area. For example where crosion rate is very low the row crops can be grown at every alternate year but on the contrast in high erodible areas or where soil erosion is being more there may be practiced only once in five or even seven years cycle.

For erosion control by growing the crops in notation system probably the most suitable crops are legumes and grasses. The main benefits credited by these crops are mentioned as under:

- Reduction of soil erosion resulting from high degree of good ground cover.
- Help to maintain or improve the status of organic content in the soil thereby contributing the soil fertility and enable to develop more stable aggregates in the soil.
- Increase in soil nitrogen resulting from nitrogen fixation associated with legume crops.

Different crops and management practices used for growing them have different effects on soil structure. The crop affect the soil structure by the activities of their root system and the amount of organic residue contents added from the roots and top of the plants. The organic contents help in arranging and stabilizing the soil particles into granules or aggregates form. This in rum to provide greater pores in the soil mass causing rapid water take in the soil. Apart from above the soil aggregates are also developed by tillage operations wetting and drying of soil freeing and thawing of soil activities of micro organism and small animal like earth worm.

Interilled crops including vegetables and grain crops usually do not have effective root systems for improving the soil structure However most of the grain crops return a considerable amount of organic matter to the soil provided that the resides after grain movable should be covered into the soil by ploughing operations. The vegetable crops rectums very little organic matter to the soil. The dense root system of grass does much to create soil structure and also helps in binding the soil aggregates together.

Tillage Practices

Tillage is defined as mechanical manipulation of soil to provide a favourable environment for good germination of seeds and crop growth to control the wees to maintain infiltration capacity and soil aeration. A well planned tillage practice provides a favourable environment, suitable for better seed germination and effective plant growth. In addition, it also protects and maintains a strong soil structure to fight against erosion.

Tillage for Soil Conservation: Tillage is an important and primary tool for conservation of the land. As per definition, its primary purpose is to provide a favourable soil environment for the plant growth which is indirectly related to the soil conservation. The effect of tillage on soil erosion is the function of its several effects on soil such as aggregation surface sealing infiltration and resistant to crosion destruction of soi8l structure either by excessive tillage or tillage operations at improper soil moisture condition tends to increase the soil erodibility, causing significant soil loss. To achieve a best result for soil conservation the following points should be considered for tillage operations.

1. Till no more than necessary
2. Till only when soil moisture is in the favourable limit and
3. Vary the depth of ploughing.

Types of Soil Conservation Tillage Practices: There are a number of modified tillage practices have been developed; each of them related to the specified objectives for providing a better soil and water - plant relations. Reducing the runoff as well as soil crosion by enhancing infiltration capacity of the soil. The important types of soil conservation tillage practices are described below:

Mulch Tillage: It is performed either by making the soil surface cloddy or mulched with the help of crop residues. Mulch tillage is happened to be an effective measure to minimize soil erosion and to conserve the moisture when it is combined with strip cropping system. This type of tillage is also practiced to utilize the crop residues as mulch and also performing farming operations simultaneously. It can be defined as a method which permits the crops to grow where all or most of the residues from previous crops are left on the soil surface. The use of mulch tillage is based on the following profits.

1. The mulch intercepts the falling raindrops over the land surface and thus dissipating their kinetic energy which result in reduction or climination of their dispersing action on the oil structure.
2. The match ultge increases the infilte capacity.
3. The obstacles caused by leave stems and roots over the field reduce the velocity of surface flow and thus controlling the sheet crosion.
4. It maintains the soil relatively cool and moist which are essential for good plant growth and
5. Increases the crop yield by developing several conducive effects on soil.

Mulching:

It is defined as the application of any plant residues or other materials ot cover the top soil surface for.

- conserving the soil moisture.
- reducing the runoff and thereby to control soil erosion.
- checking weed growth
- protecting from winter climate.
- improving the soil temperature.
- modifying the micro - environment of soil to meet the needs of seeds for their good germination and better growth of seedlings.

The mulching is known to attribute the suppression of the weed growth conservation of moisture by checking evaporation and runoff to protect the soil against erosion (mainly from wind) to increase infiltration of water to fluctuate the soil temperature to enhance mineral nutrient availability to enhance nitrification to add nutrients and organic matters derived from decomposing of residues or other materials used as mulch to preserve or improve the soil structure.

Mulching also improves the soil aeration creates better biological activates and thus to make a consequent beneficial effect on the soil fertility.

Mulching Materials:

The followings are used as mulching materials:

- Cut grasses or foliage
- Straw materials.
- Wood chips
- Saw dusts
- Papers
- Sand stones
- Glass woods
- Metal foils
- Cetto phanes
- Stones
- Plastics

Types of Mulches:

The mulches may be following types

a) Natural and
b) Synthetic
c) Petroleum
d) Conventional
e) Inorganic and
f) Organic

The natural mulches are borned by nature itself no man's effort is required.

Synthetic Mulches: Includes organic and inorganic liquids that are sprayed on the soil surface to form a thin film for controlling the various atmospheric happenings taking place over the top soil surface. The different synthetic mulching materials are as under.

- Rasins
- Asphalt emulsions
- Latex and out back asphalt
- Canvas
- Plastic and paper products
- Polythene and polyvinyl chloride (PVC).
- Bitumen emulsions.

The plastic mulches are very useful for nurseries in semi - arid and arid regions but their demerit is to have more cost and difficult to apply on large scales. The plastic mulching is also not suitable for the taller vegetations. In nurseries the dark coloured plastic mulches make the soil surface very warm during the day hours because black body absorbs greater heat from the sunlight. This type of characteristic results in lower diurnal and higher nocturnal temperature.

Petroleum Mulch: The petroleum mulches are easier to apply and also less expensive. These mulches are available inform of emulsions of asphalt in water that can be sprayed on the soil surface at ambient temperature to form a thin film in continuous form that clings to the soil but not penetrate deeply. A thin film of petrol cum substance made so is termed as mulch film. The mulch film promotes uniform and rapid seed germination and also plays a significant role for vigorous growth of seedlings. An ideal surface film is also stable against erosion sufficiently porous to allow water in the soil yet insoluble in water and resistant enough to the forces of weather causing it to last as long is necessary for permanent vegetations to cyme established.

Conventional Mulch: The mulches such as hay or straw are more effective than the petroleum mulches. These mulches not only conserve the moisture and reduce the fluctuation of soil temperature but also protect the soil from rain drop impacts and hold the excess surface water in contact with the soil so as to increase the infiltration rate and thereby reducing the runoff and soil erosion.

In skit on during day hours these mulches also absourth as resulting the surface of the mulch becomes hot and the soil on the other hand, during night hours, the mulch cools down and permitting the soil to remain warm.

The papers mulches are also counted under conventional mulch are reported to give a remarkable result. Paper mulches are observed to increase the soil temperature especially of the surface soil layers. There are several evidence to show that paper mulching proves bettle performance in improvement of soil condition besides promoting the carthwarn activity.

List at the same time the toxic elements of chemicals are coached out of the paper which has to be guarded against. The treated papers such as pitamanised have toxic effect on the plants.

Inorganic Mulches

Soil Mulch: It is an important mulch for the conditions of arid and semi - arid regions. Its application during summer and rainy seasons should be avoided.

The soil mulch is also effective to reduce the evaporation particularly where the soil is saturated as a lower depth below the top surface or the moisture content is in excess of field capacity but hot in contact with a continuous water - table. Sometimes, the soil mulches are not effective under ordinary conditions as large amount of moisture evaporation establishes a protective dry layer of the soil which if worked will cause excessive so moisture loss.

In soil mulching a loose and dried soil layer of 5 to 8 cm thick is established on the soil surface. For this purpose the land surface is ploughed for the depth as specified above and left over for sometime to get dry of the ploughed soil; after that by planking operation the tilled soils are planed in this way a layer of dry soil in loose condition is prepared over the land surface which acts as soil mulch. The soil mulch prepared so obstructs the capillary loss of water from the lower layers due to following reasons.

- Lack of close contact with moist soil lying below.
- Increase in non - capillary pore space.
- Resistance to wetting.

These effects are said to be more apparent under isothermal conditions in regards of soil and temperature. Soil mulching has beneficial effect on soil aeration also.

The formation of crust on top soil surface is also counted as soil mulching it results in clogging of the soil pores which effectively seals the lower horizons from contact with the atmosphere and also prevents the diffusion of O2 and CO2.

The soil mulching becomes more effective when it is composed of crumbs and clods of proper size which are not liable to be broken down in form of surface crust by the impact of subsequent rains. In sundy areas the soil mulch should be prepared immediately after runs and renewed after subsequent rains when they get infiltrated downward and rap the maximum amount of moisture.

The general function of mulch is to raise the soil temperature during winter season and to lower the same during summer season by allowing low heat conduction through the same during summer season by allowing low heat conduction through the mulched layer and thus to maintain the soil temperature at a uniform level. In many cases the soil mulching also plays an important role to diminish the weeds growth from the soil.

Store Mulching:It involves the spreading of stone pieces on the ground surface to conserve the moisture and also to reduce the wind crosion. It is a very old practice followed in arid zones. Soil under the stones tends to be in moist condition but the temperature of that soil is slightly higher. The soils lying below the stones Harbour small animals and involve a high nitrification. The stone mulching is also used for tapir the dews particularly in those locations where significant dew fall takes place. Central Arid Zone Research Institute Jodhpur has reported the use of rubble much which is simply combination of mall fragments of stones and bricks provides better result on moisture conservation compared to the stone mulching synthetic mulching and mulching made by straw materials.

Organic Mulches: The tree bunches twigs leaves leaf litter grasses weeds etc. and used as organic mulch to cover the soil surface.

The organic mulches are found superior than the artificial mulches in respect of conservation of moistures reduction in evaporation and runoff. By this mulch the control of evaporation is more effective particularly when rainfall takes place at frequent intervals but not found much effective when the rains are few and scattered in other words the infrequent rains and small showers may not be saved at all but for large rains which result we surface for several days with excess surface water for deep percolation these mulches may have their efficiency considerably more. Further more the light mulches are almost ineffective for controlling the evaporation because moisture conserving efficiency of mulch is inversely related to their capacity to absorb water or to extract it from the soil by capillary action. Resistant mulches do not decary sonly but last for a long time as a result they are more effective for conserving the soil moisture.

Benefits of Organic Mulches

The various advantages are listed as under:

1. Very effective in reducing the soil erosion heatuse they promote interception loss and infiltration of rain water.
2. They obstruct the import of rain drops over the ground surface and thus dissipate the corrosive power of rainfall.
3. Very effective in preserving and impacting the soil structures by criminality the crusting of soil surface and sealing of pores by runoff.
4. Organic mulches also chance the dew fall by insulating the soil the really and electrically from the atmosphere.
5. The ascorbic molehes under the condition of bighorn temperature keep the sad temperature below the highest temperature.

Water Losses and Their Control In Rainfed Agriculture

A. Use of Mulching:-

1. Mulching like grass, weeds & crop residues applied to the crop@ 5t / ha, reduce the maximum temp, of soil at 10 cm depth by 1 to 70C during monsoon season (July to Sept) by 4 to 100 C during summer season (April to June).
2. Mulches improved soil environment by way of increased moisture availability, reduction in soil temp. To optimum levels & thus higher water use & water use efficiency.

B. Tillage: - Affect soil - water relationships, aeration status, thermal characteristics & the mechanical impedance to the root penetration. Soil acts as mulch & restricts the upward movement of water to the evaporating site by reducing diffusivity gradients.

C. Use of Antitranspirants: - Define as the materials which decrease water loss from the plant leave by reducing the size or number of stomatal opening decreasing thereby the rate of diffusion of water vapour.

Two imp. Points in use of Antitranspirants:-

1. The application or an antitranspirants should restrict the water loss from the leaf surface without reducing photosynthesis, as carbon dioxide diffuse through stomata & is necessary for photosynthesis.
2. Transpiration causes cooling of the leaf surface & the use of antitranspirants should not completely stop transpiration & thus raise the leaf temperature.

Antitranspirants effective in Two ways:-

1. Through films that coat the leaf surface.
2. Chemicals that close the stomata

Examples of antitranspirants:-

1) Phenyl mercuric acetate Inhabits stomatal
2) Alkanyl succinic acids opening
3) Waxy or plastic emulsions - Film forming antitranspirants
4) White wash or kaolinite - Acts directly on wet cell walls & lower leaf temperatures & reduces vapor pressure gradient.

Effect on Photosynthesis of Antitranspirants: Antitranspirant result in size & no. of stomata's of the leaves, however, a supply of carbon - dioxide diffusion into the stomatal cavity is necessary for the occurrence of photosynthesis & it the reduction in opening results in restriction of actual photosynthesis & the yield reduction will be there.

Example of Anti -Tran spirants	***Growth retardants***
Hydroxylamine hydrochloride	Abscisic acid
Phenyl mercuric acetate (PMA)	a-NAA
Silicon	Phosphon
Cetyl alcohol	Daminozide

Stearic acid	DAMS, TIBA, MH& CMH
Chlormeunat chloride	Methyl ester
Alacholor	Alkenyl succinic adid, 2, 4 - dinitrohenol.
Desiccarts or defoliants	Crop - ripens such as
2, 4 - D	Picloram
Paraquat	Ammonium isobutylate
H2SO4	Ccc, carbanyl urea,
Na - chlorate	Bromacil, Endothal
Diquat	Bacitracin

Minimum Tillage Or No Tillage (Zero tillage) Concept In Rainfed Farming

Tillage may be defined as the practice of modifying the state of soil in order to provide conditions favourable for plant growth.

Tillage can also been defined as the mechanical manipulation of soil with certain implement or tools to provide a suitable environment for seed germination root growth, weed control, soil erosion control and moisture conservation.

In the recent past, minimum tillage concept come into existence reduing time, labour and machine operations as well as conserving moisture and reducing erosion. The modern technology of herbicides & insecticides made it possible to achieve some tillage requirements without using implements.

Any tillage practice in dry lands which does not return more than its cost by increasing yield and improving soil conditions should be eliminated. Soil need to be worked only enough to assure optimum crop production and weed control.

Aims and objectives of tillage in Rainfed farming:

1. Moisture management: - Soil configuration for in situ moisture conservation, to increase infiltration rate to increase moisture storage capacity of soil profile, to increase aeration to reduce evaporation losses through intertillage operations to provide drainage to remove excess water etc.
2. Erosion control: - contour tillage contour cultivation tillage across the slope.
3. Weed control: - check weed growth & avoid moisture competition.

4. Management of crop residues: - Mixing of trash and decomposition of crop residues retention of trash on top layers to reduce erosion.
5. Improvement of tilth: - minimize the resistance to root penetration, improve soil texture & structure etc.
6. Improvement of soil aeration: - For good growth of crop.
7. Providing food seed - soil contact.
8. Preparing fine surface for seeding operation.
9. Incorporation of manures, fertilizers and agro chemicals (weedicide & soil amendments) into the soil.
10. Insect control.
11. Temperature control for seed germination.

– Minimum Or Optimum Or Reduced Tillage

– It denotes the reduction of number of operation by planting directly after harrowing without any other intervening cultivation which are usually carried out to give a fine seed bed. Or

Minimum tillage is a method aimed at reducing tillage to the minimum necessary for ensuring a good seed bed rapid germination satisfactory stand and favourable growing condition.

- Objectives:-
    1. Reducing energy input and labour required.
    2. Conserving soil moisture and reducing erosion.
    3. Increase organic carbon, improve structure of soil, increase hydraulic conductivity of soil, increase infiltration of water.
    4. Providing optimum seedbed rather than homogenizing the entire soil surface.
    5. Keeping the field compaction to minimum.

Minimum tillage made practicable and economical because of:

1. Development of good equipments for combined tillage & sowing operations.
2. Enormous progress in chemical weed control which has reduced unnecessary many other tillage operations.
3. Minimum tillage frequently gives as good as or even better yields than conventional tillage methods.

- Advantages of minimum tillage:-

1. Increases organic carbon.
2. Improves soil structure
3. Increases hydraulic conductivity of soil.
4. Increases infiltration of soil.
5. Reduce soil compaction.

More advantages in coarse & medium soils than heavy soils.

- Disadvantage of minimum tillage:-
    1. Seed growth / intensity is increased
    2. Less decomposition of organic manures & release of nitrogen
    3. Less germination of crop seeds.
- Minimum tillage can be practiced by different Methods:-
- A) Ploughing planting: - In this method only a single trip over field is required. The tractor pulls a plough & planter simultaneously. The seed row is centred on the furrow slice. The area between the rows remains ploughed & weeds do not germinate easily.

This involves in less cost in seed - bed preparation and yields remain some to that of conventional tillage. The disadvantage in this method is that planting is slowed down & sowing is delayed beyond the optimum time.

- B) Till planting :- (Special Till planter):- It prepares seed - bed & sows two rows in one operation. The seed - bed is prepared by an implement equipped with a narrow & deep penetrating sweep a wider & shallower sweep and selection of rotary her. The strip between the rows neither is nor disturbed.
- C) Wheel track planting: - The field is ploughed as usual. The seedbed a prepared by wheels of the factor. The soil between rows remain rough & loose & absorb better moisture reduces runoff. Weeds seeds he dormant in loose soil until rainfall.

Save 40% villager cost ploughing + planting of seed shout be done at one time to avoid drying of upper soil surface.

Zero Tillage

Zero tillage refers to tillage systems in which soil disturbances is reduced to sowing generals and traffic only and where weed country must be achieved by a genital nears. It can be considered as a men extreme form of minimum tillage 2010 tillage maintains more corposiders than any other tillage soil surface and it protect the grout against wind and water evasion.

Primary tillage it commonly avoided and secondary tillage restricted to Seedbed Corporation in the row zone only. It is also known no till and is resorted to who soil are subjected to wind & water erosion Zero tilled soil are homogenous in structure with high population earthworms. Organic matter content increases due to less mineralization.

Control of weeds is the main problem in zero tillage. Incomplete weed control is the main ebstoute to the further adoption. Zero tillage a widely used in humid areas.

Erosion losses and polities are minimized by zero tillage. Zero tillage will be useful concept where than.

a. Soils are subject to wind and water erosion eg. in sloppy bare compacted soils with high gilt fine sand.
    ii. Timing of tillage operation 15 foods difficult.
    iii. Conventional tillage to not yield more.
    iv. Requirement of energy and labour too high.
    v. In medium to fine textured soils use of heavy implements can result in formation of hard puncturing wet conditions. Much more research information is needed on cartilage.
b. Concept of Minimum Tillage is Useful In Rainfed Farming
c. How the concept of minimum tillage or zero tillage is useful in Rainfed farming:
d. It has been pressed hearing the experimental findings that the conventional tillage practices to not give higher yields over the maximum tillage practices in dry lands & hence the minimum tillage concept is useful in reducing additional cost on unnecessary tillage practices. The practice of harrowing alone may serve the purpose of seedbed preparation.
e. Frequent tillage operations results in loosening the top soil layer frequently which is subjected to more soil erosion due to intense rains. The research findings in dry lands have indicated that the frequent tillage operations lead to higher soil erosion as compared to untilled or less tilled soils. Hence the minimum tillage concept is useful in Dryland framing in reducing soil erosion.

f. The crop residues left over the soil surface acts as a much and helps in minimizing the evaporation losses. These crop residues also reduce the runoff losses, thus help in soil and water conservation in dry land.

g. The organic or crop residues get incorporated in top soil layers in subsequent period and increase the organic mater content of soils increase the infiltration rate of water reduce the bulk density increase the soil aggregation reduce the compaction of top soil layer thus increasing the productivity in dry lands.

h. Frequent tillage operations in dry lands also leads to formation of hard pans in heavy soils when worked under wet conditions & hence frequent tillage operations be avoided in heavy soils of dry lands.

i. The fine textured heavy soils of Dryland posses the self cracking habit extending to the depth of one metre and thus serves the purpose of ploughing Hence such soils should not be ploughed every year. The research findings have indicated that such soils can be ploughed once in three years.

j. The problem of weed control can be avoided by using the effective herbicides for various field crops in Dryland. Thus, the tillage operations required for weeds control can be reduced under Dryland conditions.

k. The concept of zero tillage is not applicable in any kind of Agricultural system including dryfarming at this stage since sufficient research information need to be generated for its successful application.

Harvesting of Rainfall

(Water harvesting or Run off Farming in Rainfed Agriculture)

About 10 - 20 percent of the total rain goes as runoff in medium deep black soils. This also considerable soil loss by way of erosion. The extent of runoff varies with rainfall intensity and its duration land topography soil type and land use pattern. This runoff otherwise going as waste can be collected in suitable water storage structures such as farm ponds and used further for crop production. This technique of collection of runoff water during the period or excess rainfall and its further use for crop production is called "water harvesting" or "Runoff farming". Such collected water is used to provide supplemental irrigation to the crops at the most critical growth stages or during the prolonged period of drought. In water harvesting the part of land from which the water is received is called "donor area" or "water producing area" or water harvesting area or watershed area or catchments area and the area in which it is used is called as "Recipient area" or crop production area. The donor area generally is not suitable for crop production.

Method of Water Harvesting

There are three method of harvesting and recycling of runoff water.

i) Inter plot water harvesting: - In this method harvested water is directed to the crop. This method is suitable for area where rainfall is scanty (< 500 mm) and even there is difficulty of maturing a single crop. In this technique a portion of the area is cultivated & remaining area is used for harvesting water. Usually the uncultivated area is compacted or treated in such a way that runoff would be induced. Surface modification may be required to get runoff. Such method is suitable for arid regions. Runoff may be induced by using cover films (plastic or rubber) preparing hydrophobic layer (wax) compacting surface or spreading sodic soil on surface.

ii) Inter row water harvesting: - There may not be enough rain to support a crop in some areas & therefore by conserving more water in furrows and planting the crop in furrows may give some yields.

iii) Water harvesting in farm Ponds: - A portion of the excess runoff water after allowing maximum in situ moisture conservation is collected in farm ponds. As far as possible the pond should be located in the lower patches of the field to facilitate better storage and less seepage losses. The size of the farm pond should be worked out considering annual rainfall probable runoff and the catchments area. Generally, 10 to 20 per cent of the seasonal rainfall is considered as runoff in medium and deep black soils. A farm pond of 150 m3 capacity with side slopes of 1.5: 1 is sufficient for each hectare of catchments area in black soils. The farm ponds may be circular squared or rectangular. However eared or rectangular ponds are more convenient for harvesting of runoff water.

Under low rainfall situations to increase the runoff from catchments area the soil surface is treated with sodium salt betonies clay hydrophobic compounds like sodium ciliolate sodium rosinate etc. asphalt bitumen and water proofing membranes like paraffin. Some mechanical measures to increase runoff can be adopted such as land surface smoothening reducing surface depressions compacting the soil surface by rollers of spreading the clay blanket before rolling in sandy soils.

There are three important stages involved in water harvesting.

1. Collection of water in form pond.
2. Storage of water & problems
3. Applications of stored water to the crops.

Collection of Water in Form Pond: - From the parametres like annual rainfall probable runoff and area of catchments the size of the farm pond can be work out. The location of the farm pond should be such that there v. should be proper storage and facilities to whiles storage and facilities to utilize stored water for crop production e.g. if farm is located in rocky and porous part, it would difficult to use stored water for crop production. Under such circumstances water may be increased and surface area may be required to convey for long distance. As for as possible pond should be located in lower patches of the field to facility better storage and seepage losses. The size of the farm pond should be decided by the quantum of water to be stored and nature of the soil strata. If the stratum is hard, rocky then it would be desirable to have shallow pond. If the structure is clay wool in that case depth may be increased and surface area may be reduced to have minimum evaporation. If the pond is located in upper patches water can be. How gravitationally and may create problems or water logging in low lying areas if proper care is not taken.

Storage of Water Problems: - Seepage and evaporation losses of stored water are the major problems of farm ponds. Nearly 40 to 50% quantity of stored water is lost through seepage and evaporation. When the pond is with murrum strata or under neath the total losses can be to the extent of 72% out at which 80% losses are due to seepage alone. In general the seepage losses in deep black soils are low. For preventing the seepage losses in farm ponds located in coarse textured soils the sealing materials such as natural clay, saling sodic soils, bentonite bituman, soil + cement mixture stones or brick in cement mortar asphalt compounds polyethylene / rubber sheets or plastering with soil + cowdung wheat straw etc. can be used for lining the pond surfaces depending upon the easy availability and cost of the material. However compacting and lining with natural clay soil is most economical.

The evaporation losses from free water surface can be reduced by spreading the materials on water surface such as plant residues oil emulsions long chain fatty alcohols i.e. Cetyl alcohol gum mixtures polyethylene oxides. Floating blocks of wax rubber and plastic floats are more effective in controlling the evaporation to the extent of 80 percent.

Application of Stored Water to the Crops: - Since available water in the farm pond is a scare commodity its optimum use is the important consideration in entire runoff farming. In the case of application of water for crop production two considerations need to be borne in mind. First is the method of application and second the stage of crop growth.

For efficient application furrow irrigation or alternate furrow irrigation methods should be used than surface irrigation which will increase water use efficiency of stored water. When the stored water is to be used for post rainy season (Rabi) crops the water should be applied at the most critical growth stages.

For example: Rabi sorghum - When 2 irrigations are to be applied

First - stem elongation stage - 30 - 35 DAS.

Second - Flowering 65 - 70 Days. When stored water is limited and only one irrigation is possible in that case water may be applied before flowering to avoid storage losses. Here water should be stored in soil profile rather than in farm pond.

Safflower - 60 - 65 Days - Rosette stage.

Gram - 65 - 70 Days - pod development stage.

The research findings from Solapur have indicated that grain yields of Rabi sorghum safflower and gram can be increased by 100, 40 & 60% respectively by applying single irrigation at boot stage rosette stage & pod development stage respectively to above crops.

All these things discussed above should be combined into a method of management called "watershed based farming system". In this new approach the attempt is made to utilize water in all its stages and then excess water is drained out in to a farm pond connected to the field by protected grass water ways.

Methods of Controlling Runoff

A. Mechanical Methods:

1. Contour bunding
2. Graded bunding

3. Biological Bunding or live Bunds or vegetative bunding; or Vegetative barriers
4. Water shed management- inter bund management
5. Broad bed furrow
6. Vertical mulching

B. Agronomical practices:-
1. Strip cropping
2. Mulching
3. Contour cultivation
4. Planting of grasses for stabilizing bunds.
5. Intercropping
6. Sequence cropping
7. Relay cropping

Vertical Mulching: This is the practice followed in dryfarming areas for moisture conservation. The in filtration rate to black soils of dry lands is very low. In the event of high intensity rainfall much more water is lost as runoff instead of infiltrating into the soil profile. This process still accelerated under sloppy lands. Under these conditions the technique of vertical mulch has been found useful in Dryland farming.

This technique consists of digging suitable trenches across the shope and thus making more surface area available for absorption. The open trenches are filled with organic farm wastes like straw stubbles the stalks etc. which is called as filter. The filter should be resistant to decomposition and provide service for 3 - 4 years. The upper portion of filter should be 15 - 20 cm above the soil surface.

The trenun should be of 20 cm width in between two crop rows. The trench depth of 60 to 90 cm is optimum. The interval between trenches should be 4 m. the runoff water is trapped by the filter and allowed to percolate in the trenches the stored water in trenches recharge the soil profile by lateral movement of water. The findings on vertical mulching at Solapur & Mohole indicated 35 to 40% increase in grain yield of Rabi sorghum.

Vegetative or Biological Bunding: The bushes like Subabul shevri of the grasses like vitiveria i.e. khus grass are planted in between the bunds in the fields across the slope or along the average contours. The system is called as vegetative bunding or biological bunding. The grasses or the bushes are cut close to the ground periodically leaving 20 to 30 cm top portion above the ground. This above ground portion helps to arrest the surface flow of excess water. The water halts temporarily along the vegetative bunds and helps in silting of soil particles. During this time water gets some time to infiltrate into the soil. Then partially clear excess water goes up to the field bunds with non erosive velocity which is further drained into field drains. The interval between two vegetative bunds will depend on the slope of the field. However 10 - 12 m interval between two bunds is convenient for carrying out field operations.

The bushes like Subabul or shevri can also be planted at 15 - 20 m intervals across the wind direction in the fields which acts as wind breaks and useful for checking soil erosion and moisture conservation.

Effective Rainfall: From crop production point of view it is the portion of rainfall which contributes to the crop water needs is the effective rainfall. In other wounds the and of in tall watch becomes the part of consumptive use of water of a crop. An activicual farmer considers that the effective rainfall whiches that total rainfall which is useful in raising crops planted on his soil. Water which moves out of his field by surface runoff is the portion of total rainfall which is ineffective. Also the water that moves below root zone as deep percolation is ineffective. Any rainfall received after the soil has attained the field capacity up to rot zone depth is ineffective.

Factors Affecting Effective Rainfall

1. Rainfall characteristics:-
   i) High intensity of rainfall - less effective rainfall
   ii) More duration of rainfall - less effective rainfall
   iii) Well distributed rainfall - more effective rainfall
   iv) With light showers.
2. Land characteristics:-
   i) Leveled land - more effective rainfall
   ii) Sloppy land - less effective rainfall more runoff
   iii) Ploughed land - More effective rainfall
   iv) Vegetative cover - Less runoff, more effective rainfall.
3. Soil characteristics:-
   i) Infiltration rate - high infiltration rate - more effective rainfall

ii) Storage capacity - More storage capacity - more effective rainfall

(Depth of soil)

iii) Initial water content - high initial water content then less will be the effective rainfall

Crop Characteristics: More roof zone depth complete ground cover Active stage of growth - More uptake of water more will be effective rainfall.

Climate: More radition High temperature and Low temperature Relative humidity – High crop requirement and more effective rainfall

Vegetative Cover: Surface condition of soil canopy soil cover natures of roots mulches etc. affects the effectiveness of rainfall.

Drought Resistance And Characteristics Of Drought Resistant Crops And Their Varieties

Drought is a hazard to successful production. It occurs due to various combinations of the physical factor of the environment. Internal water stress in crop lands reduces their productivity. This reduction in productivity is brought by

1. Adely or prevention of crop establishment
2. Weakening or distraction of established crops
3. Predisposition of crops to insects & diseases.
4. Predisposition of crops to insects & diseases.
5. Alteration of physiological & bio - logical metabolism in plants & alternation of quality of grain forages fibre etc.

Drought:-

i) "Deficiency of available soil moisture which produces water deficits in the plant sufficient to cause a reduction in plant growth. Or

ii) "Drought is a period of inadequate or no rainfall over extended time creating soil moisture deficit and hydrological imbalances."

Classification of Drought:

On the basis of source of water availability B) On the basis of occurrence C) On the basis of media

1) Meteorological drought 2) Agril. Drought 3) Hydrological drought 4) Socio - economic drought 1) Permanent drought 2) Seasonal drought 3) Contingent drought 1) Soil through 2) Atmospheric a drought

On the Basis of Water Availability

Meteorological drought: - Ramdas (1960) defined this as actual rainfall is deficient by more than twice the mean deviation.

Indian Metrological Department (IMD) has defined meteorological drought as the situation when actual rainfall is less than 75% of the normal rainfall over an area. This is accepted principally because of its simplicity. The IMD uses two measures to define drought conditions.

i) Rainfall conditions

ii) Drought severity.

Rainfall conditions:-

i) Excess - 20% more than average of 70 - 100 yrs.

ii) Deficient - 20% less than average of 70 - 100 yrs.

iii) Deficient - 20 to 59% less than average of 70 - 100 yrs.

iv) Scanty 60% less than average of 70 - 100 yrs.

Drought severity:-

The IMD classifies droughts as follows from rainfall departures.

i) Slight drought when rainfall departure is 11 to 25% from normal rainfall.

ii) Moderate drought when rainfall departure is 26 to 50% from normal rainfall.

iii) Severe drought when rainfall departure is 50% and more from normal rainfall.

Drought years the year is considered drought when less than 75% of the normal rainfall is received.

Hydrological Drought: - Definition of hydrological drought is concentrated with the effects of dry speels on surface & sub surfaces hydrology rather than with the meteorological explanation of the event. Linsley et al. (1975) considered hydrological drought as "a period during which stream flows are inadequate to supply established used under given water management system". The frequency and severity of hydrological droughts often defined on the basis of water depletion or shortage in reserve basins, reservoirs lakes wells etc. This drought affects industry and power generation.

Agricultural Drought: Heathcoat (1974) defined agricultural drought as the shortage of water harmful to man's agril. Activities.

This is a situation resulted from inadequate rainfall when soil moisture falls short to meet the water demands of the crop during the growing period. This affects the crop growth or crop may wilt due to moisture stress resulting in yield reduction.

Socio - Economic Drought: The socio - economic effects of drought can also incorporate features of meteorological hydrological and agricultural droughts. They are usually associated with the supply & demand of some economic goods. This drought should be linked hot only to precipitation but also trends of fluctuations in demand.

On the Basis of Time of Occurrence: Droughts differ in time and period of their occurrence. As per thornithwaite the various droughts are:

Permanent Drought: This is the drought area of permanent dry arid or desert regions. Crop production due to inadequate rainfall is not possible without irrigation in these areas. Vegetation like cactus thorny shrubs, xerophytes etc are generally observed.

Seasonal Drought: In the regions with clearly defined rainy (wet) and dry climates seasonal droughts may result due to large scale seasonal circulation. This happens in monsoon area.

Contingent Drought: This results due to irregular & variability in rainfall especially in humid & sub humid regions. The occurrence of drought may coincide with critical crop growth stages resulting in severe yield reduction.

On the Basis of Medium: - Maximov (1929) has defined into 2 types.

1. Soil drought: - It is the condition when soil moisture depletes & falls short to meet the potential Evapotranspiration (PET) of crop.
2. Atmospheric drought: - This results from low humidity dry and hot winds & causes desiccations of plants. This may happen even when rainfall & moisture supply is adequate.

Drought Prone Area

A drought prone area is defined as one in which "the probability of drought year is greater than 20%

Cronic drought prone area: A cronic drought prone area is defined as one in which the probability of a drought year is greater than 40%.

The criterion described above is useful for a continuous monitoring of the monsoon season. The sum of the seasons rainfall becomes the basis for describing a region under moderate or severe drought. When more than 50% of the area in the country is affected is described as severely affected by drought & when the area of 26 - 50% of the country is affected it is described as an incidence of moderate drought.

What is Drought Resistance?

It is the ability of a plant to maintain favourable water balance and turgidity even exposed to drought conditions there by avoiding stress and its consequences. Stress avoidance due to morphological anatomical characteristics which themselves are the consequences of the physiological processes induced by drought these zerophytic characteristics are quantitative and vary according to environmental conditions.

A favourable water balance under drought conditions can be achieved by transpiration before as soon as stress is experienced. These are called "water savers" or.

Accelerating water uptake sufficiently so as to replenish the lost water called as "water spenders"

A) The mechanism for conserving water:-

Stomatal Mechanism: -Stomata of different species vary widely in their normal behaviour and range. In some species stomata remain open continuously or remain closed continuously. Many cereals open their stomata only during a short time in the early morning and remain closed during rest of the day. There is a difference in this respect between varieties of the same crop as shown by the example in two varieties of oat one is more resistant to drought open its stomata more rapidly in the early morning when moisture stress is at its minimum and photosynthesis can precede with the least loss of water (stocker 1960).

However mechanism of conserving water based on the closure of stomata will inevitable load to reduce photosynthesis and may lead to drought induced starvation injury (Leavitt, 1972).

Increased / Photosynthetic Efficiency: - On possibility for overcoming limitations on photosynthesis, imposed bicoastal closure as means for increasing resistance to loss of water by transpiration there by transpiration there by accumulations of CO2 would be at higher rate for a given stomatal opening (Hatch & stack, 1970). A number of imperfect crop plants (maize, sugarcane sorghum prose, fox tail & finger millets) (Hatch et. al. 1987) as well as certain forage species Bermuda grass (Cynodon dactyl on) Sudan grass Bahia grass (Paspalum notatum) Rhodes grass (chloris Guyana) (Murata lyama 1963) and certain A triplex sp. fixed most of CO2 into the C4 of molic and aspartic acids so called C4 dicarboxylic acid (C4) pathway.

Low Rate of Cuticular Transpiration: - The typical example is the cacturs. Thick cuticle results in low rate of transpiration.

Decreasing transpiration by a deposit of lipids layers on the surface of the leaves on exposure to moderate drought e.g. soybean (Levitt 1972).

Reduce Leaf Area: - The principal means of reducing water loss of xenomorphic plants is their ability to reduce their transpiring surface. Apart from the common means of keeping the aerial parts small perhaps the simplest form of this reduction of the transpiring surface is the sealing or of leaves at the time of water stress a characteristic phenomenon exhibited by many grasses. The rolling of leaves has been shown to reduce transpiration by almost 55 percent in semi conditions and by 75 percent in desert xerophytes (Stalfect - 1956).

Leaf Surface: - Various morphological characteristics of leaves he reduce the transpiration rate and may affect survival of plants drought conditions. Leaves with thick cuticle waxy surface and the presence of spines etc. are common and effective.

Stomatal Frequency and Location: - A smaller number of stomata retard the development of water deficits. In certain species, the stom are located in depression or cavity in the leaves which is feature can further reduce transpiration by limiting the impingement of currents.

Effect of Awns: - Awned varieties of wheat predominate in the drier at warmer regions and have been found to yield better than awnless one especially under drought conditions though there are exceptions (Gurandhacher 1963). Awns have chloroplasts stomata and so as photosynthesized. It has been found that the contribution of the away to the total dry weight matter of the kernels was 12% of that the entire plant.

B) To Improving water uptake (MC - Donough & Gauch 1959) :-

Efficient Root System: The root systems of drought resistant plants are characterized by wide variety of apparent adaptations. These responded to such predominant soil conditions as the duration of soil dryness and the depth that is normally wet. Plants become adapted to dry conditions mainly by developing an extensive root system rather that structural modification of the roots (shields - 1958). The conceres "extensive root system" includes additional growth of secondary hair roots.

High Root to Top Ratio (R/T): A high root to top ratio is very effective mean to adoption of plants to dry conditions of the growth rate of the roots considerably exceeds that of the shoots. The transpiring surface is there by reduced while root system of the individual plant obtains it's water from a large volume of soil (Simonis 1992) has shown that an increased root top ratio may actually result in greater amount of total dry matter of plants grown under dry conditions as compared a similar ones grown with full moisture.

Difference in Osmotic Potential of Plants : Levitt (1958) has calculated a difference of 0.5% in soil moisture content that includes per manual wilting could supply a plant with enough water to keep it alive for 6 days. This could mean in certain cases the difference between survival and death.

Conservation of Water Spenders to Water Stress: Because of increased water absorption water spenders are characterized by very high rate transpiration. However as soon as the absorption rate becomes insufficient to keep up with water loss the water spenders generally develop some of the characteristics of the water savers (Cevitts - 1972).

Mitigating Stress

Mitigating Stress: Adoptions a drought basis mitigating effects of stress permit the plant to maintain a high internal water potential inspite of drought conditions. They therefore able to maintain cell tartar and growth avoid direct or indirect metabolic injury due to dehydration (Levit 1972).

Drought Tolerance: When plant is actually submitted to low water potential it can show drought tolerance by either mitigating the actual stress induced by the moisture deficiencies or by showing high degree of tolerance to stresses.

High Degree Tolerance; Resistance to Dehydration: The simplest method of avoiding drought induced damage is by resisting dehydration, preferably tot he extent .of maintaining turgur and at least by avoiding cell collapse after loss of turgur (Levit 1972) retain their turgur and therefore can continue to grow when exposed to drought stress. When plants are grown in their natural environment their osmotic potentials tend to be characteristic for each ecological group.

Characteristics of Drought Resistant Plants

Early Closure or Stomata: Opening stomata for short time in easy morning & remained closed during rest of day when moisture stress is minimum with photosynthesis with the least loss of water e.g. varieties of wheat, oats.

Increased Photosynthetic Efficiency: The plant species using the pathways have a high rate or carbohydrate assimilation for given stomatal opening higher temperature & light optimum e.g. maize sorghum.

Low Rate of Cuticular Transpiration: Thick cuticle results in low rate in transpiration e.g. cactus.

Deposits of Lipid Layers: On exposure to moderate drought conditions the lipids are deposited on leaf surface which in reduction transpiration losses e.g. soybean

Reduction in Leaf Area: Rolling or curling of the leaves reduces the leaf surface exposed to sunlight thus helps in reducing the transpiration loss under stress conditions. E.g. Maize, Sorghum, grasses etc.

Waxy Leaf Surface: The leaf surface becomes waxy forming thick cuticle and develops spines on leaves which help in reducing transpiration losses. Eg. Safflower.

Stomatal Frequency and Location: Location of stomata in cavity or in depressions of leaves reduces the direct contact of stomata with wind currents & reduces the transpiration losses. In drought resistant plants, the number of stomata found more a lower leaf surface. Similarly, the number of stomata is also reduced which helps in reducing transpiration losses.

Effect of Awns: The Awned varieties of wheat barley etc. can thrive well under stress conditions as the awns contain chloroplasts & stomata & can continue photosynthetic activities even when the stomata on leaves get closed during day time.

Accelerating Water Uptake: The water uptake by plants in increased efficiently due to following plant characteristics.

i) Efficient root system:

i) Extensive root system

ii) Deeper root system

iii) Secondary root etc.

iv) Ability of roots to go towards available water

v) Ability of roots to penetrate in soil

ii) High root to top ratio:

1. Transpiring surface is reduced.
2. Water absorbing surface is increased.
3. High osmotic pressure: Under stress conditions the osmotic (Low osmotic potential) potential in roots and aboveground plant parts in reduced resulting increased water movement through soil and plant.

Nature of Varieties Suitable for Rainfed Farming

1. Varieties should have medium height with early grand growth period. e.g. rabi sorghum varieties - Selection - 3. SFV - 86, M -35 - 1.
2. Varieties should have medium tillering habit, bigger ear head size & bold grain size. E.g. bajara variety - shraddha.
3. The variety should have deep and extensive root system.
4. The variety should be of shorter duration.
5. The variety should have High Harvest Index.
6. The intercrop varieties should be of longer duration with differentiation growth habit. E.g. Red gram varieties BDN - 2, No. 148.
7. Varieties should be resistant to moisture stress.
8. Varieties should have coating either with wax or other material which prevent the loss of moisture through evaporation from stem and leaves. E.g. Rabi sorghum varieties - white glumes on stem & leaf sheath. Safflower varieties - waxy surface & spines on leaves.
9. The varieties should be photo and thermo insensitive e.g. Gr. nut variety - TAG - 24.

Package of Practices of Crops Under Rainfed Conditions

Choice of Crop and Varieties For Rainfed Agriculture: Dryland constitutes about 75 per cent of cultivatable lands in the country. The contribute about 42 per cent of off grains, almost all the coarse grains and 75 percent pulses and oilseeds. More than 90% of sorghum, Pearl millet, groundnut and pulses are grown in arid and semi arid areas.

At present, the cultivable area under Dryland agriculture in the state is 87% and only 13% area is under irrigation. After harvesting all available water resources, it is possible to bring it to 30%. It means

70% of the cultivable area will remain as Rainfed in the state. Under this situation the state will have to depend upon for its major share of food production on Dryland.

However, the Dryland agriculture suffers from two problems viz. low productivity and high in stability.

The reasons for low productivity in Rainfed area are:

1. Lack of moisture conservation practices.
2. Low rate of fertilizer use.
3. Lack of timely farm operations.
4. Improper crop planning as per land capability.
5. Inadequate efforts to increase water resources.
6. Unpredictable rainfall situations.
7. Lack of improved Technology

The adoption of improved Dryland technology will be the only answer to mitigate above situation.

Considerable research efforts are being made in the state to develop the improved the improved Dryland technologies since early seventies which will help to improve the crop production in Rainfed agriculture.

The important improved Dryland technologies are given below:

1. Selection of efficient crops and their varieties:
2. Crop planning as per length of cropping season:
3. Developing suitable varieties for dry lands.
4. Seeding time for Dryland crops
5. Timely seeding for pest avoidance
6. Planting pattern and plant densities
7. Intercropping
8. Fertilizer use in Dryland
9. Weed Management
10. Use of minimal irrigation
11. Crop planning as per land use capability
12. Crop planning for aberrant weather situation in dry lands
13. mid - season correction practices
14. Soil and water conservation practices

Improved Dryland Technologies

Selection of efficient crops and their varieties, Crop planning as per length of cropping season

1. Selection of efficient crops and their varieties:

Improved varieties and hybrids of Kharif and Rabi crops have higher moisture use efficiency as compared to local varieties. Hence improved varieties are adopted for efficient moisture use.

Kharif crops: Bajara - Shraddha (RHRBH - 8609, Saburi (RHRBH - 9824)

Sunflower - Modern, SS- 56, EC - 68414, KBSH - 11, APSH - 11)

Gr. Nut - SB X 1, K - 4 - 11, ICGS - 11, TG - 26, Koyana (B - 95)

Red gram - No. 148, BDN - 2, ICPL - 87; TT - 6, T - Vishakha - 1.

Cowpea - Konkan Sada bahar.

Soybean - MACS - 13 Pk - 472, Monetta, JS - 335.

Setaria - Arjun

Horse gram - Sina, Man (Kulthi or Hulga)

Green gram - PM - 2, S-8, Jalgaon - 781, BM - 4, THRM - 18.

Black gram - T - 9 TPU - 4, TAU - 1, TAU - 2.

Castor - Aruna, VL - 9, Girija

Kidneybean - MBS - 27 (Matki),

Grasses - Marvel - 8; Stylo, Siratro.

In general, the use of improved varieties increases the grain yields by 20 to 25 percent over local varieties. Hence sowing of these varieties should be carried out in Rainfed Agriculture.

Rabi Crops:

Rabi sorghum - M - 35 - 1 (Maldandi), Selection - 3, Swati (SPV - 504), CSV - 14 - R.

Gram - PG - 12, Vijay, Vishal, N - 59, Chaffa, Bharati,

Safflower - Bhima, Girana

2) Crop planning as per length of cropping season:

A) Cropping season with less than 20 weeks:

Single crop either in Kharif or Rabi.

Kharif - Bajara, Green gram, Gr. nut, black gram sunflower.

Rabi - Jowar, Safflower, gram.

B) Cropping season with more that 30 weeks:

Two crops with short duration Kharif crops following by 100 - 120 days rabi crops.

E.g. Bajra / Green gram - R. Jowar, Safflower, Gram.

C) Cropping season with 20 - 30 weeks:

Suitable for intercropping e.g. Pearl millet + Red gram (2: 1)

Developing Suitable Varieties For Dry Lands

The varieties or hybrids suitable for dry lands should have flowing characteristics.

i) Short duration, medium height, high yielding ability.

ii) Big ear head size with bold grains.

iii) Resistant to water stress conditions.

iv) Strong penetrating root system.

v) High Harvest Indes

Eg. Grunt JL - 24, TAG - 24, Bajra - Shraddha - Safflower - Bhima

Horse gram - Sina, Man etc.

Seeding time for dryland crops:

Proper time of seeding is important in dry lands as the let growing season is likely to be shortened. For this rainfall probabilities.

Eg. Cotton, red gram, horse gram dry seeding in 24th Meteorologist week at Solapur found optimum.

For this the off season tillage to be practiced to shorten the time aces between first rain and actual seeding time. It also helps to increases moisture.

Timely seeding of Rabi crops is also important eg. Sorgurti & Safflower - Traditional practice - end of September. Improved practice - first fortnight of September. This helps in better utilize of soil moisture and nutrients.

Timely Seeding For Pest Avoidance:

Timely seeding of Kharif crops found useful in avoiding cest incidence.

E.g. Kharif sorghum should be sown before early. July to seed shoofly incidence at seeding stage and midge fly incidence at flowering to grain formation stage.

Planting Pattern and Plant Densities

Under adequate soil moisture conditions change in planting pattern has no advantage. However, it is necessary while adopting intercropping systems to accommodate intercrop rows.

E.g. Kharif - Bajra + Tur in paired planting in 2: 1 row proportion (30 - 15cm.). Under limited soil moisture paired planting is useful during the season for efficient moisture paired planting is useful during Rabi season by efficient moisture use.

E.g. Rabi sorghum 30 - 30 - 60 cm. or 45 - 45 - 90 cm spacing. This is due to deeper & more root growth and convenience in inter culture operations.

Plant density - While deciding the plant density, the availability of stored soil moisture needs to be considered.

Gram - Low soil moisture - wider planting - 60cm.

high soil moisture - closer planting - 30 cm.

Sorghum - Low soil moisture - 5 - 10 plants / M2.

High soil moisture - 5 -10 plants / m2.

Safflower - Not affected by plant density

Bajra - 10 - 15 plants / m2 optimum.

Safflower - 1 to 1.25 Lakhs plants / ha optimum.

The optimum plant population leads to higher production per unit area.

Sr.No.	*Crop*	*Spacing (Cm)*	*Plant population in lakhs / ha*
1.	Bajra	45 x 15	1.5
2.	Groundnut	30 x 15	2.5
3.	Red gram	60 x 20	0.75
4.	Horse gram	30 x 10	3.30
5.	Moth bean	30 x 10	3.30
6.	Setaria	30 x 5	6.0
7.	Sunflower	60 x 20	1 to 1.25
8.	Gram	30 x 10	3.3

Developing suitable varieties for dry lands, Seeding time for dryland crops, timely seeding for pest avoidance and Planting pattern and plant densities

Crop Planning For Aberrant Weather Situation In Dryland

The following common weather aberrations are observed in Solapur region of dryfarming area in Maharashtra.

1) Delayed onset of monsoon.

2) Good start of monsoon followed by dry spells.

3) Good start of monsoon followed by dry spells.

3) Early withdrawal of monsoon.

4) Extended monsoon.

The following crop management practices can be followed under different aberrant weather situations.

Delayed onset of monsoon: Onset of delayed monsoon is a common feature of the dryland agriculture.

If the Kharif soils (< 45 cms) are diverted for rabi sorghum under such delayed rains, it not only affect the production of Kharif crops especially pulses and oilseeds but also results in poor production of rabi sorghum.

Hence such shallow soils must be put under Kharif crops. Substantial research efforts were made at dryland centre. Solapur in this regards and following cropping pattern has been suggested depending on the lateness of monsoon during Kharif season.

Rainfall Situation Suggested Crops Remarks

Rainfall situation	*Suggested crops*	*Remarks*
1. Normal onset of Monsoon. Rains during Ist fortnight of July	All Kharif crops Bajra, Setaria, sorghum, grunt, castor, Red gram, Horse gram, Black gram, sunflower	Adopt intercropping of Bajra + red gram in 2 : 1 preparation
2. Late onset of monsoon Rains during 2nd fortnight of July (Rains delayed by 15 days_	Setaria, Sunflower, castor, Red gram, Horse gram (delate Bajra, sorghum, gr. nut, black gram)	Intercropping of Red gram + Setaria in 2 : 2 Proportion.
3. Very late onset of monsoon Rains during Ist fortnight of august (Rains delayed b 7 y 30 days)	Sunflower, red gram, castor, castor, horse gram (delese Setaria from above set)	Intercropping sunflower + Red gram in 2 : 1 proportion
4. Very very late onset of monsoon Rains during 2nd fortnight of august (Rain delayed by 45 days)	Castor, sunflower red gram (Delete horse gram from above set.)	Red gram in 2:1 proportion
5. Extremely late onset of monsoon Rains during Ist week of September (Rains delayed by 50 days)	Rabi jawar for fodder	

i) Controlling plant population for conserving and effective are of available moisture.

ii) Checking weed growth to reduce moisture loss.

iii) Increasing interculturing to prevent evaporation.

iv) Choice of crops like red gram and castor which can sustain longer breaks.

Early withdrawal of monsoon: This situation creates two problems in rabi.

a) Sowing of Rabi crops may be suspended.

b) When crop is sown, requires moisture conservation practices such as :

 i) Reduced plant density: Rabi jawar sown at early September with 1 to 1.35 lakh / ha require 50% reduction. This plant population needs to be adjusted before plants go for their grand period of growth (30 - 35 DAS)

 ii) Use of surface mulch: Moisture can be conserved by using organic surface mulches. For this purpose apply mulch @ 5 t/ha.

 iii) Protective irrigation: If possible protective irrigation may be given. Usually, protective irrigation is given at 55 - 56 days growth. However, due to early withdrawal of monsoon, the same may be applied at 35 - 40 days growth.

 iv) Increase frequency of Intercultivation: Early stoppage of monsoon results in early cracking in soil. To prevent cracking and loss of moisture, frequency of Intercultivation may be increased. Untimely, 3 interculturing are recommended. Same can be increased to 5 or 6 which acts as a dust mulch.

 v) Stripping of leaves: Upper 3 - 4 leaves are retained & lower leaves are removed to reduce transpiration.

Extended monsoon: Such situation is rarely experienced. Double cropping is possible in medium deep soils. Sowings of Rabi crops are extended. In certain areas sorghum may be replaced by gram and what as the sorghum may suffer due to cool speel.

Climatologically Approach For Crop Planning In Rained Areas

Agriculture production in India is closely related with rainfall India. About 75% of total cropped area is rained. Crop product on area is very uncertain due to erratic behaviour of rainfall. The main for very low and highly unstable yields in these areas is unavailability adequate soil moisture occurring active growth period of crops. The moisture stress can be mitigated if it is followed by good rainfall. How prolonged stress period affects the rinal crop production. There are reasons for low productivity but primary constraint to this is lacs suitable tautology for soil and water management.

A sustainable farming system needs management strategies with respect varietals selection soil fertility programmes and pest management agricultures order to reduce the input cost and provide a sustained avel of production profit from farming.

For planning an efficient agricultural production system information weather and climate is vital and also a major resource for agriculture productivity and sustainability especially in a stressed environment considering the complete macroclimate system.

It is necessary to develop farming systems that are sensitive to climate and weather sustainable and to susceptible to degradation on account of climate.

In some areas the total rainfall is sufficient for one good crop and the some cases for two good crops in a year.

However the rainfall distribution in root profiles some times exceeds and percolation of water to deeper layer or ground water recharge takes place. Because of the uncertainties and ever present risk of droughts the farmers are generally reluctant to adopt the use of available high yielding varieties, fertilizers and other input management which will effectively conserve and utilize soil and water.

Weather plays an important role in crop production, more so in India where 75% of the cultivated area is rained. The effective cropping season in rainy season is restricted by both rainfall quantity and distribution, thereby setting limits on choice of crops, cultivars and cropping systems. For post rainy season crops grown on conserved soil moisture, it is moisture storage at sowing time that determines the choice of crops and cultivars.

Crop growth rate at different ptenophases, length of growing season, efficiency of PAR interception, efficiency of solar energy conversion, efficiency in biomass conversion are explained to quantity the impact of evicemental parameters on growth and development of crops. However in arid and semi - arid areas water use plays a dominant role in crop production.

In many arid and semi arid areas crop production problems follow a familiar sequence:

i) Unfavourable crop growth environment:-

i) Limited choice of crops and cultivars, particularly in water deficit environments and aberrant weather situation.

iii) Low crop intensity.

iii) Low and onstable productivity.

Water deficits are responsible for low and unstable crop yields in both arid and semi arid areas. In addition, environment stresses / or nutrient stress may make the water deficit environment more unfavourable for the growth. The crops and cultivars currently popular in dryland areas are in necessarily the most stable and efficient in terms of moisture use. Many the existing cultivars of sorghum, pearl millet, pigeon pea, groundnut sunflower, castor and other crops are not adapted to rainfall pattern when they are grown for effective cropping season. The usually experience drogue stress at the most critical stages of their life cycle, which leads to low at unecrowic yield. In order to achieve yield stability it is necessary to ground crops and cultivars with water requirement patterns that watch the effective growing season.

Climatic Variability In Rainfed Agriculture

In India about 70% of cultivated area is rained which contribute to about 40% of the country's food production. Appa Rao and Bhide, 1980). The main climatic parameters controlling crop growth are rainfall followed the temperature, adiation day length, humidity and wind speed. The inter seasonal and inter a variability in these climatic parameters play a major role in deciding the proper agronomic management options and subsequent realization of the yield. The major characteristics associated with the south west rainfall are high variability in its distribution in time and space in its cnset and withdrawal and frequent and prolonged dry spells as a result of break in the monsoon. The uncertainty in rainfall received ouring the growth of crop become a major limitation factor in deciding about the final yield.

In the day sew arid areas the men annual rainfall exceeds potential

These areas suffer from one or more or a combination cf factors such as moisture deficits. Limit of nutrients soil version and physical conditions, resulting in low infiltrations and poor crop establishment and subsequently larger yield gap. Thus crop yields in semi - arid India have remained low and variable because of aberrant weather and soil related constraint such as poor management of soil fertility and rain water.

Time and Length of Growing Season - Tropical Regions

In tropical regions where low temperature does not limit growth, the time and length of the growing season for sorghum is determined by the seasonal precipitation pattern. Kassam et.al. (1978) and Kassam (1979) used precipitation data and computations of potential Evapotranspiration (PET) (Thorntwaite, 1948) to determine the growing seasons for crops in tropical Africa.

The first day (a) wren the normal precipitation becomes equal to or greater than half the normal PET is the beginning of the growing season and earliest planting time. The last day of the growing season (c) is the day when the normal daily precipitation becomes less than half normal PET plus time required to evaporate 100 mm of stored moisture from the period when precipitation exceeds PET.

The sorghum growing seasons for different tropical areas in eastern end western Mexico. The growing season at Villahermosa which receives 1902 mm. of rainfall is 333 days. At Apatzingan (716 mm rainfall) the season is only 125 days. A study by Kassm (1979) shows that a creed relationship exists between the amount of annual rainfall and the length as growing season in Africa.

Table: *Growing season as related to precipitation at two locations to tropical Mexico.*

Location	***Rainfall***	***Growing season***
Villahermosa (170 59 N 920 55 W)	1902	15-Apr
Apatzingan 190 05 N 1020 15 W)	716	4 Nov.

Growing Period or Moisture Availability Periods

Length of the growing period is defined as the period during which the availability of moisture in the root zone of a crop is adequate to meet the water needs. Because the amount and distribution of rainfall varies considerably from year to year so does the effective growing period. The length also depends on the type of soil interacting with a given quantity on rainfall. In areas receiving rainfall for 2 months, the growing season may the 8 days in a coarse textured soils or 100 days in soils of clayey or day texture. Similarly in areas with 5 rainy months, the growing from 180 to 210 days depending upon soil texture and moisture tolding capacity (Virman - 1991).

Therefore, at a given location, the amount and distribution of rainfall, moisture storage capacity and the rate of Evapotranspiration determines the length and characteristics of the growing period.

Soil moisture reserves have the ability a extend the growing period by as much as one to three months spending upon the soil texture and depth.

The soil moisture availability period determines effective cropping season. Based on the analysis of long term data in the arid and semi - arid areas of India effective cropping period have been delineated for a number of locations.

In arid areas, the effective cropping season is normally 11 - 17 weeks, which restricts the choice of crops and limits the farmer to a single crop in the rainy season.

In semi - arid regions, the effective cropping season is normally longer (22 - 32 weeks) with exceptions of 8 weeks in Bellary (Karnataka) and 17 weeks in Bijapur (Karnataka) regions. Rainy season crops are grown on shallow to medium Vertisols at Bijapur, while post - rainy season

Table: *Effective cropping season at various locations in arid and semi arid areas of India.*

Zone	*Location*	*Growing Season (Weeks)*	*Monsoon (23 - 39 mm)*	*Rainfall (mm) Post - monsoon*
Arid	Jodhpur	11	353	8
	Hissar	13	395	19
	Anantapur	13	305	149
	Rajkot	17	572	36
Semi	Hyderabad	22	603	108
Arid	Bangalore	32	400	226
	Bijapur	17	381	130
	Solapur	23	494	101

(Source: Single and Sebba Reddy, 1988)

crops are commonly grown on deep Vertisols at Bellary. The rainfall pattern and soil depth together determines the choice or crops and cropping systems. On shallow to medium Alfisols and related soils, only single season cropping mostly during the rainy season is possible. The amount of pre-monscore ram received in May determines whether or not double cropping is possible on demo Alfisols.

The water balance for different dryland research stations of SAI. The India has been calculated (Sing - 1993) and water availability periods haven been worked cut. The water availability period ranges between as low as 105 days at Bijapur and Bejary and as high as 210 days at Bangalore.

Table: *Water availability period at different dryland research centres of the SAT in India.*

Soil type & centre	Rainfall (mm) Total	Dependable	Water availability Period Day	Range (mm)	Total duration (weeks)
1. Vertisols and related black soils :					
Rajkot	674	532	134	25 - 44	20
Udaipur	661	572	164	25 - 48	24
Akola	878	702	196	25 - 52	28
Indore	1054	858	196	25 - 52	28
Jhansi	999	809	196	25 - 52	28
Solapur	743	584	168	23 - 46	24
Bijapur	537	434	105	33 - 47	15
Bellary	519	387	105	33 - 47	15
Kovilpatti	724	622	135	39 - 05	19

(Source: Singh, 1993)

Rainfall at 75% probability on long term basis Based on the information of water availability seriods potential cropping system have been suggested for different situations. The selection of efficient cropping system is also influenced by selection of suitable crops and their cultivers passed on duration and water use efficiency besides.

Climate, Soils and Socio Economic Factors (Singh, 1993)

Farmers have to adjust the cropping systems and crop management practices to the limitations imposed by the environment. The farming system which they have practiced has been developed by experience of generations without proper knowledge of agro climatic conditions, effective cropping pattern and schedule of supplemental irrigation can not be planned. For this study of moisture availability index (M.A.I.) is very important. The cropping patterns are basically dependent on moisture availability index (M.A.I.). Hargreaves (1971) defined M.A.I. as there ratio of assured rainfall expected at 75% probability and estimated potential Evapotranspiration for the concerned period. However, Thormathwite and Mather (1965) calculated MAI by using water balance equation. Bistroni (1980) has defined M.A.I. as:

$$MAI = \frac{AE}{PE}$$

Whereses,

AE = Actual Evapotranspiration

PE = Potential Evapotranspiration

For determining actual, Evapotranspiration (AE) following two conditions nave to be considered.

1) If P > PE, then PE = PE

2) IF P < PE, then ae = p + s

Where,

P = Precipitation

S = Change in soil moisture

Rainfall distribution over the country is highly erratic and tooth in time and space and there by Moisture Availability Index (M.A.I.) also becomes very underlain.

Moisture Availability Index the prime factor for especially in tropics waries both in time and space. The MAI on the basis of average monthly rainfall (Roman and Marth, 1971 and planning was done. However, in such system the monthly MAI values where truly representative as month is a longer period for planning any operation,. Moreover, If there are dry spells in between causing failure, the monthly MAI may not represent it. Hence, there is need is weekly MAI for agriculture planning. For planning of majority of crops the weekly MAI values would be most available.

Raskar (1994) determined crop growing period on the cases of availability index (MAI) of different rainguage stations of Pune, Ahmednagar districts of Maharashtra. In rainfall zone 1 to 3 of scarcity of these districts, the crop growing period at 0.3 MAI ranged between 18 a 19 weeks and at 0.5 MAI between 13 and 17 weeks during that season and 0.3 MAI between 9 and 11 weeks and at 0.5 MAI between 7 and 9 weeks during rabi season. The dry spells of 3 to 6 weeks duration were observed.

Table: *Weekly moisture availability period at 0.3 and 0.5 MAI in different soil type in various rainfall zone of scarcity tract of Pune and Ahmednagar districts.*

Rainfall	***Kharif***						***Rabi***					
Zone	***Shallow***		***Medium***		***Deep***		***Shallow***		***Medium***		***Deep***	
	0.3	0.5	0.3	0.5	0.3	0.5	0.3	0.5	0.3	0.5	0.3	0.5
1*	18	13	18	13	18	13	10	9	10	9	10	9
2**	19	16	19	16	19	16	11	7	11	8	11	8
3***	19	17	19	17	19	17	9	7	9	7	9	7

(Source: Rasker 1994)

Main of 5 iccation ** Mean of 8 locations *** Mean of locations at different locations. Drop planning on the basis of MAI was suggested sorghum grown on conserved soil moisture is sown in 39 the week but analysis suggests that it is quite possible to sow the crop well ahead of the present practice i.e. by 39 mm, preferably by 35 mm itself, when the assured amount of soil moisture is available for sowing as indicated by MAI more than 0.5. This will help the crop to have sufficient moisture in latter active growth stages such as elongation, flowering and grain filling.

Importance of Rainfall Distribution

The average annual or seasonal rainfall at a place does not give sufficient information regarding its capacity to support crop production. Rainfall distribution pattern is the most important. This is illustrated by taking examples of Hyderabad and Solapur. These two locations are 500 km apart but have very similar amount of total rainfall (742 mm at Solapur, 764 mm at Hyderabad). At both the places, more than 75% of total rainfall is received September. However the distribution of rainfall at Solapur is highly erratic during Kharif. In comparison, Hyderabad has more dependable rainfall distribution pattern and thus more favourable for Kharif cropping than Solapur (Virmani et.al. 1982).

Field experiments at these two locations have shown that at Hyderabad, it is rossible to produce more that 500 kg grain ha-1 on deep Vertisols by adopting pigeon pea + maize intercropping C- maize - chickpea sequence cropping under good agronomic management. At Solapur Kharif cropping is undependable for a long duration crops but short duration crops of pearl millet sunflower or grain legume in Kharif followed by rabi Sorghum or gram is successful.

Some Practical Application of Weather Data

Prostitution of rainfall within a season and the frequency of occurrence of dry speels of different durations can help to select the optimum time of planting and fertilizer application. If the probable dates of such saving irrigation, under taking or wittroiming of a nitrogen top dressing car also be made with greater confidence.

Temperature and Humidity: Temperature as the dominate factor controlling rate of development. The diurnal temperature cycle is more important than either the seasonal cycle or random effects of weather in the SAT (Monteeith, 1977) Even more important for plant growth processes and incidence of pests and diseases are the effects of microclimate. Humidity is also an important agro climatic factor because it is a major determinant of evaporation and incidence of pests and diseases. Nair et al. (1995) studied the influence of meteorological parameters on the incidence of shoot fly on Kharif sorghum order field conditions. They closer that egg laying and population dynamics stowed highly significant corretation with meteorological parameters like temperature, relative humidity, bright moisture tours and rainfall intensity while dead heart formation was not correlated

The incidence of powdery mildew on grape appeared early and more vigorously where relative humidity was high due to irrigation at shorter intervals (Chavan et. al., 1995). The temperature in the range of 11.8 to 32.42 C and relative humdity) 58.4% favoured the development of powdery mildew. Wereas, temperature below 8.6 and above 34.09 C and relative humidity below 47.4% showed zero rate of multiplication indicating that disease did not multiply though, it existed. The incidence and multiplication was rapid in the months of December and Jawary when the climate was cool and humid.

Management Option In Relation To Weather Adjustment

While in irrigated condition timeliness of irrigation is important for higher production, a number of options such as choice of suitable crops and varieties, alternate crop strategies, mid season correction, crop life saving measures, alternate land use systems etc. are to be adopted in rained agriculture to adjust to aberrant weather conditions.

Cropping Systems: Suitable cropping systems aiming to adjust or reduce the intra - seasonal impact of climatic variability should be based on inter - seasonal variability of rainfall the water deficiency and the length and characteristics of growing season.

Intercropping: In arid and semi - arid region of Hyderabad, the length of growing season varies from 15 to 30 weeks with variation in rainfall and soil moisture storage capacity. In soils having medium available water storage capacity (150 mm in the soil profile) a crop with a 19 weeks growing season is likely to have adequate moisture only coce in 4 years. Under such conditions.

Table: *Length of the growing season (week) for three soil conditions Hyderabad.*

Rainfall Probability	***Growing Season (weeks)***		
	Available water storage capacity (mm)		
	Low (50 mm)	Medium (150 mm)	High (300 mm)
Mean	18	21	26
75%	15	19	23
25%	20	24	30

(***Source:*** Virmani - 1989)

From seed germinating rains 25 June to put of season (time when profile moisture reduces AET / PET ratio of Actual evaportanspiration to potential evapotranspration to 0.5)

**Low:* shallow Altisol : medium : shallow to medium - deep Vertisols: high deep vertisoils. intercropping of a short duration sorghum (105 - 110 days) with a long our at lon pigeonpa (150 - 180 days) yielded a land equivalent ratio of 1966. Pan and willey 1981). Thus, in case of random variability of rainfall, intercropping increases crop yield as well as provides stability. Based on 89 sorghum pigeonpea intercrop experiments conducted in diverse environments. It was been observed that on an average intercropping yielded equivalent of 90 mm solv sorghum yield and about 25% of sole pigeonpea yield.

Choice of the right intercropping will depend of the awards distribution pattern and soil moisture storage. In areas having be ental rainfall in the early part of the growing season drought tolent be (pigeonpea) may be useful. If rainfall is undertain is the later part of the growing season, then intercrop spould be storter in than the base crop Considering the high valves of pulses and outseeds and their soil restoring ability, these crops shouln find a place in any probulable intercropping programme.

Intercropping of pearl millet + pigeonpea (2:1) and sunflower + pigeonpea (2: 1) on medium deep soils are ideally suited for dryland conditions (Jadhav et.al., 1991) due to large difference in maturity periods of component crops which useful for harvesting the natural resources like solar rediation, soil moisture and nutrients more efficiently.

Table: *Mean grain yield kg. ha-1, LER and gross monetary returns (Rs. ha-1) as influenced by treatements (Pooled data of 3 years : 1985 - 1988).*

Treatments	***Grain yield***		***LER***	***Gross monetary***
	Main crop	***Intercrop***		***returns***
Pearl millet + pigeonpea	1082	250	1.58	3564
Sunilower + pigeonpea	639	224	1.42	4049
Sole pearl millet	1053	-	1.00	2350
Soil sunflower	700	-	1.00	3222
Sole pigeonpes	442	-	1.00	2339
S.E.	-	-	-	390
C.D. (p 0.05)	-	-	-	1125

(Source: Jadhav et. al., 1991)

Proportionate Cropping

In this system land area allocated to crops of different growing duration on the basis of long term probabilities of soil moisture.

Research conducted at OCS Haryana Agricultural University, Hisar allocating 40% of land to guar U 20 d durational. 40% to pearl millet (70 days) and 20% to mungbeen (50 days) enabled to harvest all three crops in good rainfall years and at least two crops in all but severe drought years (Virmani, 1989).

Thus, proportionate cropping can help to decrease the risk of loss and increase over all productivity.

Moisture Deliverinacy and Cropping System: In covered and rained regions because of combined effect of variable rainfall.

High Evapotranspiration rates and poor water tolding capacity of soils crops are often exposed to suhoptimal moisture availability our mg our or more capital plenological stages of crop growth.

The adverse effect of moisture deficiency can be minimized by choosing crop or crop varieties with duration appropriately fitting to moisture availability periods.

Table: *Potential cropping systems in relation to rainfall and soil type.*

Rainfall (mm)	*Soil type*	*Water availability*	*Potential clopping*
		period (week)	*systems*
350 - 600	Alfisols and shallow	20	Single Water if crop
	Vertisols		
350 - 600	Aridisols and Entisols	20	Single crop enther
			water if or ran
350 - 600	Deep Vertisols	20	Single ran crop
600 - 750	Alfisols and entisols	20 - 30	Intercrcpping
750 - 900	Entisols, deep Vertisols	20	Double cropping with
	Alfisols, Inceptisols		monitoring
0 - 900	Entisols, deep Vertisols.	20	Double cropping
	deep Inceptisols		

(*Source:* Katyal et al., 1994)

Contingent Crop Planning: Weather aberrations are important features of deyland agriculture. One season seldom matches with another.

As such every year poses a new situation. It if therefore, not enough to develop the tectrology for normal weather conditions but strategy needs to be developed for aberrant weather suitability.

In fact these aberrations are the part of the propping situations.

Analysis of weather data and crop conditions at Solapur reveals that out of a years, 3 years are normal and ?

Years are abserval in scar rate tract. Water delayed onset of monsoon 20 - 30% of the total cropped area team.

Table: *Contingent crop planning on shallow and medium deep sols (45 cm depth) of scarcity zone of Maharashtra.*

On set of monsoon	***Crops suggested***
(Fortnight)	
June II	Pearl millet, sorghum, pigeanpea, sunflower, green gram black gram, proundnut, horse gram, kindney bean, castor, etc. and intercroppings.
July I	Pearl millet, sorghum, horsegram, kidney bean, pigeon pea, groundnut, sunflower, castor etc. & intercroppings
July II	Pigeonpea, sunflower, horse gram, kindney bean castor
August I	Pigeonpea, sunflower, horsegram, castor.
August II	Pigeonpea, sunflower, castor
September I	Rabi sorghum (M - 35 - 1) for fodder or selection 3 for grain

(***Source:*** Patil et al., 1981)

monson for shallow to medium deep soils of scarcity tract of Maharashtra.

Management Strategies Under Different Draught Conditions

Depending upon time and intensity of moisture stress, management strategies have to be adopted. Moisture stress periods are usually classified be (i) parly season moisture stress (ii) mid - season moisture stress (iii) terminal stress.

Early Season Stress

Early season moisture stress occurs due to the failure of rains after sowing of crops /cropping systems or delayed start of rainy season. If produced dey spell occur immediately after sowing the seedlings may water.

Mid reserve stress may occur due to the period of monsoon after mateblislwent of crops. Under these situations adoption of frequent interculturing aerations use of organic mulches, lop dressing of outrage after rained of moisture steps formation of dead furrow red such dry spells in the same season, the likelihood is quite high Bhavanisagar (77%), while it is les at Solapur (15% once in 7 years) and the chance at Hyderabad is zero.

Extension advisors, field agronomists and district agriculture officials can use such information by using the data from their regional metrological stations in order to time tune the blanket recommendations to suit particular conditions.

***Table:** Probabilities of selected events at three locations.*

Event or opportunities	*Probability (%)*		
	Solapur	*Hyderabad*	*Bhavanisagar*
Possibility of at least one off season	38	54	92
tillage in May			
Success in early panting	77	92	15
Occurrence of one 21 day dry spell during			
June to September rainy season	92	54	100
Occurrence of three 21 day dry speel in			
one rainy season	15	0	77
Occurrence of three 14 - day dry spell in			
the rainy season	62	38	100

(Source: Huda et al., 1988)

Management Strategies Under Different Draught Conditions

Depending upon time and intensity of moisture stress, management strategies have to be adopted. Moisture stress periods are usually classified be (i) parly season moisture stress (ii) mid - season moisture stress (iii) terminal stress.

Early Season Stress

Early season moisture stress occurs due to the failure of rains after sowing of crops /cropping systems or delayed start of rainy season. If produced dey spell occur immediately after sowing the seedlings may water.

Mid reserve stress may occur due to the period of monsoon after mateblislwent of crops.

Under these situations adoption of frequent interculturing aerations use of organic mulches, lop dressing of outrage after rained of moisture steps formation of dead furrow red such dry spells in the same season, the likelihood is quite high Bhavanisagar (77%), while it is les at Solapur (15% once in 7 years) and the chance at Hyderabad is zero.

Extension advisors, field agronomists and district agriculture officials can use such information by using the data from their regional metrological stations in order to time tune the blanket recommendations to suit particular conditions.

Table: *Probabilities of selected events at three locations.*

Event or opportunities	*Probability (%)*		
	Solapur	*Hyderabad*	*Bhavanisagar*
Possibility of at least one off season	38	54	92
tillage in May			
Success in early panting	77	92	15
Occurrence of one 21 day dry spell during			
June to September rainy season	92	54	100
Occurrence of three 21 day dry speel in			
one rainy season	15	0	77
Occurrence of three 14 - day dry spell in			
the rainy season	62	38	100

(***Source:*** Huda et al., 1988)

Interaction of Sowing Date and Climatic Variability

The effect of sowing wheat at different dates was simulated for all 138 locations (Das and Karla, 1995). It was apparent that as potential yield increased. The reduction in yield per day delay in sowing also increased.

In general, the yield decrease was between 0.25% and 0.75% of potential yield when the latter was less than 4% hari, between 0.5 to 1.00% for yield potential between A and 6% hart and between 0.75 to 1.00% for yield potentials greater than 6. It is interesting first irrespective of potential yield, a few location showed a small yield reduction (less than 0.25%) with delayed sowing. For New Delhi environment, the maximum grain yield was obtained for sowings done between 1 and 15 November.

The effect of varying amounts of post sown irrigation for wheat was tested for New Delhi environment with WIGROWS model in relation to the seasonal climatic variability (using 20 years runs from 1971 - 93). The amount of moisture at the time of sowing was assumed to be 75% of the field capacity. The amount of water applied at each irrigation was assumed to be 60 mm. Nitrogen application at the rate of 150 kg hart was applied at the time of sowing. The result showed increased yield with increase in number of post sown irrigations and stabilizing beyond three irrigation. Variability width seemed to be dependent of the amount of post - son water received. It is decreased with increase in post - son water received. It is decreased with increase in post - son water received by the crop, indicating the reduced effect of climatic variability on yields under increment moisture availability condition. The variability index values ranged from about 0.59 up to one irrigation to around 0.135 (beyond for irrigations) with intermediate values of 0.42 (Two irrigations) and 0.20 Three

Fertilizer Use In Rained Farming, Levels and Methods Or Fertilizer Application

One of the important management practices to increase the crop production. In dry lands is the use of fertilizers. Fertilizer is next important component to moisture in dry lands. Therefore, it is said that the soils of dryland are not only thirsty but hungry also. Soils of this zone are generally poor in nitrogen content (Total nitrogen content 0.03 to 0.05%) and they respond to nitrogen application. Available phosphate status is low to medium (10 to 30 kg P2O5/ha.) Response to phosphate application is noticed only during Kharif on sallow soils, which are poor in phosphate. However, there has not response to the phosphate on medium deep and deep soils (rabi soils). The reason may be that these soils are medium in phosphate and the drop requirements like jawar (M - 35 - 1) are not very high. Potash content of the soils are quite high (300 to 750 kg available K2O/ha) than usually required for dry land crops. Potash is abundant and it would be difficult to expect any response to dry crops due to application of potash.

Experiment ants on use of inorganic fertilizers are in progress since 1957 at Solapur and other locations. The fertilizer use has done much progress in dry lands.

Kharif Crops:

1. ***Bajra:*** Bajra is an important Kharif crop. During last 15 to 20 years hybrid varieties are becoming popular in dryland areas. Bajra crop responded to nitrogen application very well. Response to P2O5 was very small. On shallow and malin help, there is likely to get reapions P2O5.It is recommended to apply 50 kg N + 25 kg P2O5/ ha. The whole does of N & P should be applied at sowing. There is no interaction between N and P. If money is short P2O5 application may be deleted.
2. ***Setaria:*** Setaria is a Kharif crop in dry lands for shallow and medium deep soils. It is particularly suited under delayed sowing conditions. From the results of experiments on Setaria it is observed that nitrogen application helped to increase production significantly and substantially. In case of P2O5 the response was very small. It is therefore, recommended to apply 50 kgN to Setaria crop at sowing.
3. ***Rabi Jowar:*** Rabi Jowar is the most important and major crop of the region. Earlier work on the fertilizer requirement

revealed that there was only response to nitrogen application and as such 25 kg N for medium deep soils and 50 kg N for deep about 7 to 8 kg. The experiments were conducted to see the efficiency of fertilizer by applying part of nitrogen by foliar sprays. But in this experiment it was observed that foliar application and no advantage over soil application. With the use of new dry farming technology such as high yielding varieties and early September sowing, experiment on nitrogen requirements was started during 1973 - 74. The results of this experiment showed that in early sowing, the nitrogen utilization was proper which resulted in better response to the fertilizer. As high as 17 kg. grain and 36 kg. Fodder is prepuce per kg. of nitrogen applied. The optimum level of nitro in for different varieties tried was around 85 kg/in

Effect of Despite Application on Rabi Jowar:

Studies of varieties conducted on the phosphate requirement of Rabi Jowar (M - 35 - 1) and it was observed that the phosphate application did not affect the grain production. Lack of response to phosphate may be attributed to two reasons. Firsts the soils on which the crop was grown contained medium level of phosphate (20 kg P2O5/ ha). Secondly the requirement of the phosphate for M - 35 - 1 variety may be very low which might be supplied through native phosphate alone. On an average, M - 35 - 1 removes 25 to 28 kg P2O5/ha.

Safflower: Safflower is another important rabi crop in dry farming zone. It is, usually taken as mix crop. Studies at Solapur proved that mixing safflower along with jawar is harmiture. As a result, it has been recommended to cultivate this crop at a sole crop. This crop was found to respond extremely well to nitrogen application. Response to phosphate is small, erratic and noticed only in shallow soils in some years. Response to nitrogen is very good under good soil moisture conditions. Application of 50 kg N/ha is sufficient to get good and economic returns.

Methods of Fertilizer Application

For dry land crops fertilizers are usually applied at the sowing time. For drilling the seeds and fertilizers together Jyoti ferti - seed - drill can be used. This implement can be drawn by a pair of bullocks. It is suitable for all crops except groundnut. The row to row spacing can be adjusted from 22.5 cm to 45 cm. The fertilizers are drilled 5 cm. away at proper depth developed by providing a traditional bowl rand tubes on locally available seed drills. Fertilizer needs to be placed at 10 cm. Depth and as close as possible to the seed row. The fertilizer is placed about 2 to 4 cm. deeper than seed in the same row. Locally used 3 or 4 countered seed drill can be converted to ferti - seed - drill by local carpenters.

***Table:** Manure and fertilizer application schedule for dry land crops (Kharif).*

Crop		***Manure T/ha***	***Fertilizers***			***Time of application***
			N	P	K	
1		2	3	4	5	6
1.	Kharif Jowar (Local, improved)	6-7.5	50	25	-	Whole quantity of N,P, applied at sowing.
2.	Rainfed Kharif Hybrids of Jowar with assured rainfall.	6 - 7.5	75	62	62	1/2 N-at sowing 1/2 N at top dressing. Full dose of P and K and sowing.
3.	Maize Rainfed with assured rainfall.	12-15	90	40	40	1/2 N at sowing, 1/2 N at Knee high stage. Whole of P2O5 and K2o at sowing
4.	Pearl millets (Hybrids)	5-6	50	25	-	Whole of N and P at sowing.
5.	Pearl millet (Local and improved varieties)	5-6	25	25	-	Whole of N and P at sowing
6.	Setaria (Rala)	3-5	50	--	--	Whole quantity of N at sowing.
7.	Hill millet	5-6	50	25	25	Whole quantity of N,P,K, at sowing.
8.	Groundnut	5	12.5	25	-	Whole quantity of N and P at sowing.
9.	Sunflower	5-6	25	25	-	--do--
10.	Niger	5	25	25	-	--do--
11.		5	25	25	-	--do--
12.		5	25	25	-	--do--
13.		5	12.5	25	-	--do--
14.		5	12.5	25	-	--do--
15.	Kidney bean	5	12.5	25	-	
16.	Red gram	5	25	50	-	

***Table:** Manures and fertilizers application schedule for Rabi crops*

1.		2	3	4	5	6
1.	Rabi Jowar on deep soils	6-8	50	25	--	Entire quantity of N & P at sowing.
2.	Rabi Jowar on medium deep soils	6-8	25	12/2	--	Entire quantity of N & P at sowing.
3.	Safflower	5-6	50	25	--	Entire dose of N & P at sowing.
4.	Gram	6-7	12.5	25	--	--do--
5.	Linseed	3-5	30	15	--	--do--
6.	Sunflower	3-5	50	25	--	--do--

8

Concept of Watershed Management

Out of the 20.1 million hectares of cultivable land in the state about 17.5 million hectares are under rained agriculture. At presents nearly 13 percent of the cultivated area is irrigated. After harvesting all the available water resources at the most 30 per cent of the cultivated area can be brought under irrigation. Thus, 70 per cent area would remain as rained in the state. The crop production under rained agriculture is most unstable due to inadequate, uncertain and ill - distributed rainfall. Besides. Non adoption of improved agricultural practices in rained agriculture has further deteriorated the socio - economic status of the farmers. The major food grain requirement of the state is net from the rained areas which contain mostly all grains pulses and oilseeds while in many areas heavy soils are utilized only for rabi cropping. Crop production under rained agriculture is mostly subsistence oriented producing food grains for home consumption including cash crops and fodder for livestock. The cropping patterns vary according to soils, climate, and farmer preference and to a limited extent market demands. Sorghum, cotton, pearl millet, groundnut, pigeon pea, green gram, black gram, sunflower, wheat, gram and safflower are among the important crops grown in rained agriculture. However, the productivity of the crops is extremely low due to improper crop management practices including land management treatments adopted in rained agriculture.

Most of the community lands and privately owned marginal lands which are unsuitable for arable farming remain uncultivated and serve as grazing grounds for village livestock and source of fuel supply. The Government lands are normally located at higher elevations and are badly eroded and deprived off any vegetation. The communal grazing lands are also severely denuded and eroded leaving thin vegetative cover. As a result, these areas serve as an origin of erosion.

It is estimated that out of 30.6 million hectares of the geographical area of the state, nearly 13.8 million hectares suffers from moderate to heavy soil erosion. The detailed crosion studies made in Solapur district indicated that the percentage of deep soils (depth > 45 cm) came down from 46 to 29 in a period of 75 years indicating the severity of erosion hazard.

From the studies, it is observed that the soil loss was in the range of 60 to 90 tonnes / ha per annum.

With this rate is estimated that 20 cm of fertile soil may be lost within a pan of 24 years. However, under natural conditions of weathering process the process the formatting of 1 cm top soil layer will require more than 100 years.

Agricultural development of such rained areas has remained neglected compared to irrigated agriculture. The integrated development efforts in these areas initiated in most parts in the country under the name. "Water shed Development in Rainfed Areas" since 1984 - 85.

Definition of Watershed

i) Watershed is an area above a given drainage point on a stream that contributes water to the flow at that point.

ii) Watershed is a natural unit draining runoff water to common point of outlet.

iii) The watershed is geohydrological unit or a piece of land that drains at common point. Catchments basin or drainage basin are synonymous of watershed.

Broad Objectives of Watershed Development

In general, the watershed development fulfills the following objectives.

1. To bring about increased productivity.
2. To make yields less subject to the effect of erratic rains.
3. To improve resource conservation (soil & water) and land use.
4. To create additional employment potential for the small / marginal farmers and agricultural labourers.

Principles or objectives of watershed management:

1. Utilizing the land according to its capability.
2. Putting adequate vegetal cover on the soil during the rainy season.
3. Conserving as much water as possible at the place where it falls. i.e. In situ conservation of rain water.
4. Draining out excess water with a safe velocity and diverting it to storage ponds avoiding situation hazards and store it for further sue for supplemental irrigation during stress periods.
5. Avoiding gully formation and putting checks at suitable intervals to control soil erosion and recharge ground water.
6. Maximizing productivity per unit area, per unit time and per unit of water.
7. Increasing cropping intensity and land equivalent ratio through intercropping and sequence cropping.
8. Safe utilization of marginal lands through alternate land use system such as horticulture, Agro forestry, silvipasture etc.
9. Ensuring sustainability of the eco - system benefiting the man - animal - animal - plant - land, water complex in the water complex in the watershed.
10. Maximum the combined income from the inter related and dynamic crop - livestock - tree - labour complex over years.
11. Stabilizing total income and cut down risks during aberrant water situation.
12. Improving infrastructural facilities with regards to storage, transportation and marketing.
13. Improving the socio - economic status of the farmers.

Classification Of Watershed

Classification of watershed:

1. Macro watershed: 400 to 2000 ha.
2. Micro watershed: Less than 400 ha.

Agricultural watersheds:

i) Sub watershed: 10,000 to 50,000 ha.

ii)Multiwatershed : 1000 to 10,000 ha.

iii) Micro watershed: 100 to 1000 ha.

iv) Miniwatershed: 1 to 100 ha.

Water shed Development Concept:

Watershed development refers to the conservation, regeneration and judicious utilization of all the resources viz. land, water, vegetation, animal and human within a particular watershed in integrated manner. Watershed development seeks to bring about an optimum equilibrium in the ecosystem between natural resources man and animals.

In the past, watershed development programme was aimed at mainly on the treatment of catchments for preventing situation in reservoirs. 8Xthe aim of the watershed development in present contest is quite different. It includes the improvement in productivity of dry lands through the components like

i) Crop management

ii) Soil and moisture conservation

iii) Water harvesting &

iv) Alternate land use system.

In the past full potential of new technology could not be exploited as these components were implemented in piece meal Now, this is possible due to implementation of all there components in integrated manner in each hydrological unit of watershed.

Components of watershed development:

Following are the general items of watershed development which are required to be excuted in the catchments area depending upon the prevailing situation.

1. Soil and land management
   - i) Interceptor drains.
   - iii) Graded bunding
   - iv) Bench terracing
   - v) Interbund vegetative barriers
   - vi) Grass Waterways.
   - vii) Improvement of ill drained soils
   - viii) Nala training / improvement.
2. Water Harvesting Structures:
   - i) Nala bunding
   - ii) Farm ponds
   - iii) Percolation tanks

iv) Minor irrigation tanks
v) Stop dams in nalas
vi) Underground diaphragms.

3. Afforestation cum pasture development for rural energy and forage for animals:
   i) On private marginal and culturable waste lands.
   ii) On community and Government forest lands.
4. Agricultural development:
   i) Selection of crops and their varieties suitable for local soil and climatic situation.
   ii) Adoption of appropriate cropping system.
   iii) Contour farming.
   iv) Strip cropping.
   v) Mulching and crop residue management.
   vi) Adoption of alternate land use system depending on land capability such as Alley Cropping, Agro - horticulture, silvipastural management, dryland horticulture, tree farming, and pasture management.

Significant Gains From Watershed Development Programme

Soil and Moisture Conservation: Soil and moisture conservation is the basic need in rained agriculture. Top soil is the most fertile part of the soil profile.

This layer is lost due to erosion causing decrease in yield. Agronomic and mechanical measures for soil and moisture conservation are adopted in the watershed such as contour farming, strip cropping, mixed cropping, inter - cropping, contour / graded bunding, vegetative barriers etc.

Increase in Water Storage: Due to construction of surface water storage structures like minor irrigation tanks, percolation tanks, nala bunds, farm ponds etc.

the excess runoff water is collected in these storage structures which in turn is used either for supplement irrigation for field crops, horticultural crops or for drinking water to animals. Thus, additional area can be brought under irrigation.

Increase in Number of Wells: Due to considerable improvement in ground water recharge, the numbers of dugout wells or tube wells are increased. The farmer can apply protective irrigation to various field crops whenever necessary. Thus the area under well irrigation is increased.

Increase in Cropping Intensity: Due to increase in water resources and adoption of appropriate crop management practices, and area under double cropping is increased, which results in increasing cropping intensity.

Increase in Fertilizer Use: Due to increase in water potential and moisture conservation measures, the fertilizer use by the farmers is increased.

Improvement in crop production and productivity: Adoption of vegetative and mechanical conservator measures, results in considerable reduction in soil, water and nutrient losses from the watershed area. Further adoption of improved crop management practices results in appreciable increase in crop productivity and total crop production from these areas.

Animal and Milk Production: Appropriate management of marginal lands with productive grasses and pastures, the total forage resources are increased which reflects in increasing animal component resulting increase in meat and milk production.

Increase in Afforestation and Alternate Land Use: For producing fuel, fodder and timber, alternate land use programme is implemented in watersheds. Dryland horticultural species in addition to fuel and fodder tree species have shown promise in the watersheds.

Employment Generation and Increase in Per Capita Income: Due to optimization of available resources, there is increase in employment generation to farm families throughout the year. Due to overall increase in production and productivity in the entire watershed, there is considerable increase in per capita income.

Cropping Patterns

Cropping Pattern: The selection of crops and their varieties is to be made depending on the soil and rain fail situation in the rained areas. The photo insensitive crops and varieties with shorter duration should be chosen to escape drought of different intensities. There are wide variations, location to location in water availability periods in dryland areas. Thus depending upon water availability following are the different crops and cropping patterns to suit different climatic situations.

For rained areas:

Monoculture	*Scarcity zone*	*Pearl millet, red gram, green gram, black*	
		gram, Horse gram, groundnut	
		Rabi : Jowar Safflower	
	Assured rainfall	Cotton, sorghum, red gram, black gram,	
		green gram, soybean, sunflower	
Double cropping	Scarcity zone	Kharif crops Mung / Urid Mung / Urid Sunflower Bajra Bajra	Rabi crops Safflower Jowar Gram Gram Safflower
	Assured rainfall zone	Paddy Soybean Mung/Urid Mung/Urid Sunflower	Gram Safflower Jowar Safflower Gram
	Irrigated areas	Jowar Jowar Maize Grunt Grunt	Wheat Gram Wheat Jowar Sunflower

Stable intercropping systems for rained areas:

Scarcity zone Bajra + Tur in 2: 1 row proportion

Assured rain Sorghum + Mung / Urid in 2: 1 row proportion fall zone.

Cotton + Mung / Urid in 1: 1 row proportion

Cotton + Tur in 8: 2 row proportion

Sorghum + Tur in 2: 1 row proportion.

Tur + Mung / Urid in 1: 3 row proportion.

Grassland or Pasture Management

Most of the marginal lands are not able to sustain arable crops particularly during the drought years. Such lands can be developed into dependable pastures by following soil and water conservation measures like contour trenches and contour furrows. Controlled grazing may also help in building the forage resource.

At times, native pastures are stocked with low productive and less palatable species. These pastures lack legume component, thus, making the pasture lands nutritionally deficient. Artificial renovation of such pastures is likely to provide forage of good quality as well as sufficient quantity. In rained areas, different legumes from the genera Dolichos, Leucaena, Clitoria, Cassia and Stylosanthes have been found to do well with or without grasses like Cenchrus ciliaris. But Stylosanthes has been found to be excellent in all situations with regard to persistence, nutritive value and palatability. Different grasses from the genera Dichanthium, Cenchrus, Lasiurus, Chloris, Urochloa, Panicum, and Pennisetum etc. have been observed doing well. Cenchrus ciliaris has been found to be good in most of the situations.

The pastures are easily established if they are seeded at the beginning of rainy season.

Seeds of Cenchrus ciliaris @ 1.0 Kg. Stylosanthes hamata @ 4.0 Kg and Stylosanthes scabra @ 1.0 Kg per hectare may be used as seed moistures. The seed moisture may be broadcasted on a drizzling day. After that, light raking of the soil may improve germination chances considerably.

Research investigations have revealed that application of 20 - 25 Kg N increases dry matter yield of grass species considerably.

Similarly, 30 - 40 Kg. of P205 gives good response of legume component.

For the establishment of pasture as well as for getting increased forage production the access of livestock to pastures should be controlled so that grazing pressure could be minimized.

Planning and Implementation of Watershed Management Programme

The planning and implementation of watershed management programme should be carried out in systematic way with the active participation of farmers including constitution of co-operative watershed management societies.

During implementation of the programme, following guidelines may followed.

1. The implementation programme should start from the ridge line of the watershed to the valley, not on piecemeal basis in isolated patches.
2. Development of arable and non - arable lands should be done together.
3. Forest, pasture, cultivable land and water lands should be treated as inter linked units of hydrological entity. The condition of all lands has to be improved to meet the demands of increasing man and animal population.
4. Essentially, all developmental activities are to be carried out on watershed basis. Whole watershed area needs to be covered, may be in planned phases.

The following points to be considered while preparing the master plan for watershed development.

Out Line of Master Plan of Watershed:

For preparing of master plans of the watershed, specific formats are prescribed by the Central Research Institute for Dry Land Agril., Hyderabad. The out line for preparation of master plan for development of dry lands on watershed basis are given below -

I. Introductions: General description regarding aims & objectives of the watershed approach for Dry land Agril.

II. Characteristics of watershed:

1. General information i.e. the general description of its characteristics name of watershed, general special problems on use of natural resources like soil & water.
2. Climate: Annual of monthly rainfall in mm & no. of rainy days from the harvest, rainguage station, special problems of watershed.
3. Soils: Geological features: type of soils, series, physical & chemical properties; soil survey map & report.
4. Natural vegetation: General description of the type of vegetation & the level of management.
5. Present land use & capability of classification: Detailed land capability classification under each category. Area under each class of soil Existing cropping pattern through systematic survey of the farm.
6. Socio - Economic condition: Existing land holding pattern, irrigation facilities; available draft power; labour; credit & other facilities like marketing, transport, roads etc.

III. Analysis of Problems & Potentials:

1. Existing level of crop management & reasons for non adoption of technology.
2. Existing level of Erosion control measures & suggested control measures.
3. Present level of main water use efficiency & methods of insitu moisture conservation.

IV. Improved Technology:

1. Proposed land use: Management of practices of the alternate land use system.

2. Crop management: Description of the main features of the new technology i.e. use of different production inputs, proposals to tackle aberrant weather situations.
3. Soil & water management: Specification of the engineering measures e.g. diversion drains, bunds, terracing etc.
4. Pasture management: Silvi - Horti - Agril. Pasture development details.

V. Schedule of operation: Defining the sequence of operation keeping in view with weather & available resources.

VI. Cost benefit Analysis & budget :

a) Cost benefit ratio for each component.

b) Description of subsidy pattern under each item of work.

c) Year wise - budget.

d) Likely overall benefits from the plan.

VI. Supplemental information:

a) Soil survey reports.
b) Summery of formers survey report.
c) Design; Drawing & details of major structure.
d) Agencies involved & their responsibilities.

VIII. Maps

a) Soil survey reports
b) Land capability classification
c) Existing land use
d) Existing measures for erosion control, rills, gullies etc.
e) Proposed land use
f) Contour map
g) Proposed soil & water management measures.

IX) Provision of staff, farmers training, monitoring & evaluation of the project should also be made in the master plan.

X) Involvement of farmers in planning & executions of programme through formation of co-operative watershed development societies at village levels be considered.

Plant Population, Distribution Pattern and Weed Control In Rainfed Agriculture

Cropping pattern in dryland in dependent on quantity and distribution of rainfall, soil type and its depth. In general, cropping intensity in dryland is only 100 percent.

It is observed in the farmers field that plant density is low ranging 50 to 60 thousand / ha in dryland. It is possible to increase density of population with early sowing and use of fertilizers.

In fact, in may be said that in case of rabi sorghum it would be difficult to get higher production without change in plant density to higher side to about one lakh per ha is desirable for variety M - 35. In case of high yielding varieties it is advisable to go in still higher side. In case of safflower there is remarkable adjusting capacity to plant density.

However it is desirable to have plant density between 50 to 100 thousands ha. It is advisable to adjust plant density towards higher side from practical point gives.

Plant geometrical studies revealed that paired planting for Rabi jawar found suitable for maintaining plant density. In case of sunflower planting in square or reciangular had little advantage.

A plant density of about 74000/ha with 60 x 22.5 cm spacing is desirable. In subnormal season alternate plant could be removed to reduce density to half is desirable. Density of 1 lakh / ha (30 x 30 cm) give same yield but 1000 grain weight gets reduced. The object of keeping optimum plant population is to get higher production & grain / fodder / ha.

Crop Planning For Aberrant Weather

Crop production in dryland suffers from in stability due to aberrant weather condition from time to time. Delayed monsoon results in non sowing of traditional kharif crops which accounts for nearly 25 to 30% of the total area under crops.

So also early withdrawal of monsoon interferes sowing of Rabi a crop which is main constraint of crop production in the region. Similarly, breaks in monsoon also after crop production adversely year 1972 was the lowest rainfall year which resulted in total failure of both Kharif and rabi crops under dry lands.

Mid - Season correction: Crop planning under aberrant weather condition in dry land.

Sr.No.	*Nature of rainfall*	*Crops to be grown*
1.	Delayed on set of Monsoon	
2.	Rains during July & sowing of kharif crops by end of July or early August.	Setaria (Arjun) Red gram (No. 148) Sunflower (EC 68414), caster () Horse gram (Mans, Sinha)
3.	Rains during August & Sowing up to end of August	Red gram (No. 148) Sunflower (EC 68414) Caster (Aruna)
4.	Rains during late August & sowing up to Early Sept.	Castor (Aruna). Jowar for fodder
5.	Good onset of monsoon	Sowing of all kharif crops Common situation usually one or two dry spells are noticed
6.	If dry spell exceeds tow weeks	Corrective measures a) Control of plant population b) Checking weed growth c) Increasing interculturing serious situation in drought prone area
7.	Early withdrawal of monsoon	a) Reducing of plant population from lakh to 50 thousand as in case of rabi Jowar in 35 -1 before grand growth. b) Use of surface mulch c) Protective irrigation 30 - 45 days drought. d) Increase frequently of inter culturing e) Stripping of leaves
8.	Extended monsoon	It is rarely experienced a) Sowing of grown and wheat instead of rabi sorghum. b) Double cropping would be possible in medium deep soils. c) Postponement of sowing of rabi crops.

Weed Control In Rainfed Agriculture

Weed cause considerable damage to the crops in general and in dryland particular. The weeds complete with crop plants in respect of moisture and nutrients. The moisture as such is already in short supply in dryland. In fact one of the measures of moisture conservation is to control weeds. The extent damage varies from 35 to 97% Pretillage perations usually deep ploughing reduces weed intensity control of annual weeds is not problem in rabi soils but control of perennial weeds like. Hanjalic, Kunda is the problem in dry lands weeds are found in patches. Weeds in kharif lands are the problem. The cost involved in weed control is not likely to be compensated by the production. The major composition of the weed flora is celosia (Kurdu) comeliness (Kena) and cyanotis (Echaka).

Weed control methods:

1. ***Mechanical weed control:*** Hand weeding or hoeing carried out at 30 to 35 days of sowing at grand growth period. Competition is serious during first 30 days period. During this

period plant growth is slow and there is set back to crop growth affecting tillers, panicle size and poor stand in general.

2. Chemical weed control: Under certain situations like shortage of labour, inaccessibility of fields due to rains and mechanical weed control is with great difficulty the chemical weed control has place in dry lands. In this regards reemergence of application of weedicide has a place. Use Atrazine @ 0.5 kg al / ha as a re emergence and 2.4 - D as post emergence.

Crop Rotation and Its Factors and Advantages

Growing of set of crops in a regular succession over a same piece of land (field with) in a specific period of time.

In crop rotation soil improving crops should be rotated in time over the entire farm in a regular sequence as permissible by soil, climatic and economic factors. In general cropping intensity in dryland is only 100 per cent.

At few places on partial lands occasionally two crops are taken in favourable season (Monoculture is the rule in dryland agriculture) Increasing the cropping intensity is one of the methods for increasing crop production.

Cropping intensity is increased by sequence cropping and double cropping but intercropping may also prove effective measure for increasing production per unit area.

Factors to be considered for planning of crop rotation:

1. Soil type crop and its duration.
2. Livestock on the farm
3. Occurrence of pests and diseases
4. Price and availability of Agricultural produce
5. Cost of labour.

Advantages of crop rotation:

1. Crop rotation maintains and improves soil fertility.
2. Prevent - build up of pests, weeds & soil diseases.
3. Control of soil erosion.
4. Ensures balanced programme of work through out the year.
5. Prevent or limit periods of peak (requirements of irrigation water)
6. Conserve moisture from one season to next.

Characteristics of good rotation:

1. It should be adoptable to the existing soil climate and economical factor.
2. It should be based on proper land utilization.
3. It should contain a sufficient number of soil improving crops to maintain and build up organic matter content of the soil.
4. It should provide sufficient fodder for live stock reared on farm.
5. It should be so arranged so as to make economy in production and labour utilization.
6. It should be so arranged as to help in control of weeds, plant diseases and pests.
7. It should provide maximum area under most profitable cash crop adopted in the area.

Effect of crop rotation so soil:

1. On runoff and soil loss: Crop rotation of Bajra - red gram or groundnut recorded minimum runoff and soil less (82 to 90%) followed by Bajra red gram - horse gram.
2. On bio - logical yield: Legumes cereals or cereals legumes rotations are not only beneficial for runoff but also increase biological yields.
3. Use of crop rotations according to soil moisture:
   a) Kharif season: (Shallow and poor moisture retention capacity soils.)

Crop: Bajra, Sorghum, pulses, groundnut followed by follow.

   b) Rabi season: (Medium to deep soils fairly good moisture retention capacity soils)

Crop: Sorghum, safflower, gram are rotted with Kharif Bajra sorghum etc.

Monoculture growing of a crop on the same piece of land year after year is known as monoculture or single crop system. Fallow - Jowar

(R) Or safflower fallow in rotation. In scarcity areas only two crops are taken in three years as against one crop every year. Experiments at Dry farming Research Station Shown the variability in benefits of fallow in rotation in increasing the yield of crop in the succeeding year.

Intercrop System: By following intercropping system risk is reduced (shared) cropping intensity is increased. Crop selected for intercropping (intercrop) should not compete for moisture quick growing and short duration. The medium soils, depth uptown 45 cm do not provide sufficient moisture to support two crops in sequence (double cropping) even in normal year.

These soils are therefore ideal for intercropping. A base crop and intercrop should have different duration of life and growth rhythms. At the same time crops should be cooperative. Bajra + Red gram are ideally suited for this purpose. Red gram gets of benefits from Sept showers and gives high yields. In the events of failure of later rains in Sept bajara already sown gives good yields. In the even of failure of early rains red gram compensates the production. In normal and above normal seasons both these crops boost up total production. For very shallow soils (up to 20 cm depth) Gross like Marvels planted 60 cm apart and established and horse gram is sown as intercrop was found to be most profitable. Intercropping is uneconomic and undesirable during Rabi season because Rabi crops are cultivated mainly on receding soil moisture and thus, it creates Competition for moisture. Gram and Safflower consume more moisture during early period there will be moisture stress at ear head emergence for Rabi sorghum resulting in low yields, Sequence or double cropping. In normal your (normal rainfall) there is possibility of two cops in dry land area giving increased production ranging from 100 to 300 percent over single cropping.

At Solapur seq. Viz.

Green gram / black (gram (k) - Jowar (R) and Bajra (K) - Gram (R) was found beneficial -

Mixed cropping is extensively followed for Kharif crops. Usually 4 - 5 crops are mixed. The proportion of crops varies from place to place. Seeds of bajara, red gram, horse gram, moth bean and Sesamum are mixed. The practice is common because of shortage of labour for sowing separate rows. For Rabi season sorghum safflower is a common mixture. Usually 12 - 15 rows of sorghum followed by there rows of safflower. The proportion of rows changes as per the sowing implements in practice.

Crop Mixture and Its Advantages

It is similar to inter cropping the difference that crops are either broadcasted seeds are mixed and sown or grown as mixture with in a row.

Types:

1. Cereals - legumes
2. Cereals - oilseeds
3. Fibre crops - oilseeds
4. Fibre crops - cereals

A) For Vidarbha - Khandesh tract

Jowar - black gram

Bajra - kidney beam / Green gram

Cotton 10 - 15 rows - Red gram 2 lines.

Deccan hemp - Sesamum - seeds mixed

B) For Deccan Dists:

Bajra - 5 - 6 rows - Red gram of row

Bajra - 2 - 1 rows - one row of tur.

R. Jowar - 8 rows - Safflower 4 rows

Sunflower - 2 rows - one row of tur.

Advantages:

1. To utilize available space and nutrients to the maximum extent.
2. To secure daily requirements like pulses and oilseeds.
3. To safeguard against hazards of weather, diseases and pests.
4. To provide balanced cattle feed.
5. To avail distribution of labour through out the year.
6. To get handy installments of cash returns.

Limitation: In Rabi this system is uneconomical as rabi crops are grown on recording moisture.

The growth rhythmus and duration of life cycle of the mixture is different. In this main crop get harvested earlier than mixed crop by which the mix crop produces high yield with benefit of September showers. Bajra + red gram where the duration of life cycle of bajara is less than that of red gram.

Inter Cropping and Its Advantages

Intercropping: Growing of two or more crops simultaneously on the same piece of land (field). There is a crop intensification in both time and space dimensions. There is intercrop competition during all or part of crop growth.

Type of intercropping:

1. Mixed intercropping
2. Row intercropping
2. Strip intercropping
4. Relay intercropping

Definitions of Intercropping system:

1. Mixed Intercropping: Growing two or more crops simultaneously with no district row arrangement.
2. Row Intercropping: Growing two or more crops simultaneously where one or more crops are planted in rows.
3. Strip Intercropping: Growing soil conserving and soil depleting crops in alternate strips running perpendicular to the slope of the land or to the direction of prevailing winds for the purpose of reducing errosion.
4. Relay Intercropping: Seeding planting two or more succeeding crops after flowering and before the harvest of the standing crop.

Advantages:

1. Intercropping gives higher income per unit area than sole cropping.
2. It acts as an insurance against failure of crop in abnormal year.
3. Intercrops maintain soil fertility as the nutrient uptake is made from both layers.
4. Reduce soil runoff.

Limitations: Intercropping system is uneconomical and undesirable during rabi.

Crops to be considered for intercropping.

A) Kharif crops:

1. Medium black soils:

a) Pearl millet + Red gram 2: 1

b) Pearl millet + Horse gram / Kidney bean / cow pea Inter row of pearl millet.

3. Soils up to 20 cm depth

a) Pearl millet + red gram (30 - 60 - 30 cm)

B) Rabi crops:

Safflower + Gram (2: 1)

D) Fodder for milch animals: Sorghum bajara + Cowpea or horse gram or kidney bean.

In rained areas of Maharashtra:

1. Sorghum / pearl millet / cotton + red gram / black gram or kidney bean or cowpea or groundnut.
2. Groundnut + Sunflower.

Cotton + soybean, cotton + Black gram

Safflower + gram

How Intercropping Economizes Water Use: Selection of intercrop is one the basis of duration of crop and growth rethyms. The short duration crop gets harvested and long duration crop gets the benefits of September showers and produces more yields. E.g. Pearl millet + red grams.

LER: It is observed that under the rainfall situation deviating from 2090 to 50 percent, the intercropping system of bajara + red gram is more stable as cowpea to the pure crops tried. On an average, the land equivalent ratio (LER) comes to 77 percent compared to pure crop.

Alternate Land Use Like Agro Forestry Horticulture

Pasture, Diversified Farming Systems In Rainfed Agriculture

Crop production on dry lands in general and marginal rained lands in particular results in low and unstable and often times in low and unstable, and often times, uneconomic yields, Marginal lands because of poor management are often subjected to the processes of degradation. It is estimated that nearly 70 m hectare out of a total 100 m ha under Rainfed cultivation are facing some kind of land degradation or the other. These marginal lands are not able to sustain arable crops particularly during the drought years. The Govt. of India is deeply concerned about the improvement of degraded / marginal lands. We have, therefore to think of developing some alternate land use systems for these lands.

The alternate land use systems are surer means of stabilizing both productivity of dryland and incomes of dryland farmers, besides generating more complement potential.

Day by day demand for food fodder and fuel is growing, which could be solved by selecting suitable land use system. One of the areas of research for Rainfed agriculture is improvement of degraded, marginal and sub - marginal lands by introduction of suitable. Alternate Land Use systems like alley cropping lay farming tree farming, dryland horticulture etc.

Alternate land use systems not only help in generating much needed off season employment in monocropped dryland but also minimize risk tallies off - season rains which may otherwise go waste as runoff prevent degradation of soils and restore balance in the ecosystem.

Alley – Cropping and Its Advantages

Alley cropping is a system in which food crops are grown in alleys formed by hedge rows of trees or shrubs.

The essential feature of the system is that hedge rows are cut back at planting and kept pruned during cropping to prevent shading and to reduce competition with food crops.

Advantages:

1. It provides higher total biomass per unit area than arable crops alone.
2. It utilizes off - season precipitation which otherwise would go waste,.
3. It provides green fodder during the lean period of fodder availability.
4. It provides additional employment opportunities during the off season.
5. When planted along the contours on a sloppy land, it provides a barrier to run off water holds the silt and conserves moisture. Or

Alley cropping is a farming system in which arable crops are grown in alleys formed by trees or shrubs established mainly to hasten soil fertility restoration and enhance soil productivity.

Objectives of Alley - cropping:

The main objective of alley cropping is to get green and palatable fodder from hedge rows in the dry season and produce reasonable quantum of grain and Stover in the alleys during the rainy / cropping season.

Alley - cropping - a version of agro - forestry system, could meet the multiple requirements of food, fodder, fuel and fertilizers etc.

Three Versions (Types) of Alley cropping:

Three versions of alley cropping system, based on different objectives are.

1. Forage alley cropping.
2. Forage - cum - mulch alley cropping
3. Forage - cum - pole alley cropping

In all the three systems crops are grown in the alleys and forage is obtained from lopping of hedge rows. Two components from an essential part of the system these are at the hedge rows b) the crop grown in the alley.

Need - Based Alternate Land use System And Its Advantages

1. Among the several needs of a farmer Food always remains the first priority item, although fodder requirement is more as compared to food. Some of the need - based alternate land use systems matching the land. Capability classes are discussed below Alternate land use systems.

Sr.No.	*Food (Arable Land) II and III*	*Fodder (Non - arable land) IV and V*	*Fuel / Timber / Fiber (Marginal degraded land) VI and VII*
1.	Alley cropping Agro horticulture	Horti - pastoral silvi pastoral	Tree farming Timber cumfibre (TIMFIB)
2.	Agro - horticulture	Silvi - pastoral	
3.	Intercropping with NFTs	Ley farming	
		Pasture management	

2. Ley Farming: a rotation of arable crops requiring annual cultivation and artificial pasture occupying field for two years or longer.
3. A rotation is a cropping system in which two or more crops are grown in a fixed sequence. If the rotation includes a period of pasture (a lay) which is used for grazing and conservation the system is sometimes called "Alternate husbandry" or mixed farming. The term Ley - farming denotes a system where a farm or a group of fields is cropped entirely with leys which are reseeded at regular intervals some people described any cropping system which includes leys as "Lay Farming".

Types of Ley Farming:

Unregulated Ley Systems: These are characterized by natural fallow vegetation of various grass species a certain amount of bust growth on the pasture, community grazing and lack of pasture management all of which make such systems often more short term fallow systems.

Regulated Ley System:Individual grazing fencing pasture management and rotational use of the grassland are the usuall characteristics of regulated ley system.

Advantages of Ley Farming:

1. It helps in soil conservation, improvement in structure and fertility. It acts as a self fertility regentrating system especially with respect to nitrogen.
2. With lay farming system the other important requirement I the farmer i.e. fodder for his cattle in addition to his food is easily met.
3. For rained farming it is a low risk system as they need not invest on the costly fertilizer input for the food grain crops but by the pasture legume.
4. Labour economy during the years under ley is yet another advantage. Some of the weeds which will be vigorous during the arable cropping years would be suppressed and eliminated and as such the weeding requirements of the crop would be reduced.
5. No tillage during the ley years would have other advantages like no compaction due to farm machinery more of earthworm and soil microbial activity. It will also have the other benefits like better soil sanitation less hibernation of pests and diseases of the crop plants.

Example of Ley Farming: If a farmer has say 4 hectares land in which he want to grow sorghum and caston then he can plan four year rotation as follows.

Unit	***Year 1***	***Year 2***	***Year 3***	***Year 4***
A	Stylo	Stylo	Sorghum	Castor
B	Stylo	Sorghum	Castor	Stylo
C	Sorghum	Castor	Stylo	Stylo
D	Castor	Stylo	Stylo	Sorghum

Three Alternate Land Use In Dryland Ecosystems

Consistent with the policy of conservation and desirability of preserving the integrity of the ecosystem, the alternate land use systems could be classified into

i) agro forestry
ii) pastoralism and
iii) tourism which includes wild life.

Agro Forestry Systems of Land Use: Agro forestry is a collective term for a land use system in which woody perennials (trees and / or shrubs) are eliberately mixed on the same land management unit as crop and / or animals either in some forms of spatial arrangement or in time sequence. An ideal agro forestry system should result in a sustainable increase in overall production using management practices compatible with social cultural and economic of the local population.

In agro forestry land use systems, there are three basic sets of elements of components that are managed by man viz. tree, the herb (agricultural crops including pasture species) and the animal. This leads to a simple classification of agro forestry systems as given below.

Agri silvi cultural	- Crops and trees including shrubs / vines / tress.
Silvi pastorl	- Pastures / animals and trees.
Agro silvopastoral	- Crops / pastures / animals / trees
Agrihorticulture	- Crops / fruits species
Silvi - horticulture	- Trees / fruit species
Silvi Horti pastoral	- Trees / fruit species / animal / pastures.

Agri Silvi System

Agri silvi cultural system could be practiced in areas where wood lands can be created. The planting consists of both annual crops and perennial trees. This type of approach is most commonly observed in the cultivated areas. The perennial tree species are planted in a single row or multiple of rows in a strip at a interspaces distance of 15 - 40 m between two strips. The interspaces are utilized for growing annual / seasonal crops. The preference of choice for tree selection may lie in Acacia spp. Azadirachta indica, Dalbergia sissoo; Eucalyptus spp; Casuarinas spp; Albizzia spp; and Leucaena leucocephala, prosopis spp and caliandra spp. Growing of perennial tree species on bunds / strips may also act as wind breaks in areas where high wind velocity is a problem resulting in wind erosion and desiccation of soil moisture. Certain tree species offer the possibility of providing at least a portion of optimum crop nutrients by natural leaf drop or by lopping for purposes of green leaf manuring. This includes both fixed nitrogen as well as other nutrients recycled from the deeper soil depth. This is especially true with Acacia albida which when matures (after 3 years) is said to be deciduous in the Kharif season. As such it offers less competition for light and moisture at the time when crops need them most.

In India agriculture and forestry have co - existed for many years in close proximity. Agro - forestry systems of land use are not new to our rich heritage.

Farmers from time immemorial have been growing useful tree species with agricultural crops which used to supply fodder, fuel and small timber for himself and his live stock.

The best examples available are ; growing of Prosopis cineraria (Khajri) with agricultural crops in Rajasthan and in black soils of Northern Karnataka and the other parts of the country. The other practices then prevailing where growing of perennial tree species such as Acacia's neem mango tamarind on farm boundaries.

Agro forestry offers a good scope for more efficient use of land; water, other natural and human resources. The main advantages of this system would be:

i) Progressive land improvement by providing vegetation cover and brief to bring about soil and water conservation and production of organic material for enrichment of soil.

ii) Recycling of plant nutrients is possible due to roots of perennial tree species penetrating deep into the soil and absorbing the plant nutrients and depositing on the soil surface through leaf litter.

iii) Perennial tree species are photosysnthetically active through out the year and hence produce large quantities of biomass.

iv) Legumes plants may provide for fixation of atmospheric nitrogen and thereby enrich the site conditions and

v) Provide employment opportunities to the people.

Structural Basis of Classification of Agro Forestry Systems: Structure of the system can be defined in terms its components (constituents) and the expected roles (functions) of each (manifested in terms of outputs). It is not only the nature of components that is important but also their arrangement.

Silvi - Horti / agro - Horti System: The concept of silvi - Horti / agro - Horti or combination of agricultural crops, perennial tree and fruit species could profitably be adopted in both arable and non arable marginal and sub - marginal lands.

Semi wild but useful fruit species such as cashew, Ber, annonaceas fruit species, Chiranji, phalsa carrisa, mango, sapota, guava, tamarind and jack fruit trees are planted in regular strips or inter planted with silvi component. In areas receiving higher rainfall of 1000 m and above coconut can be planted as in being practiced in coastal Karnataka and Kerala.

In the agro Horti - silvi system of land use the distance between the two horticultural plants within the strip may be quite apart to avoid competition. The inters trip space between the two horticultural plants can be used for planting fast growing economic silvi - cultural species such as leucaena casuarinas, Dalbergia teak and albizza.

The tree plants are cut for wood after 4 - 6 years so that the competion could be minimized. The idea behind planting tree species as an intercrop with horticultural plants is to obtain biomass production before horticultural plants attain full growth, and later to obtain fodder or green manure material by frequent cutting and create thicker vegetation for better soil dwater conservation.

Many plant species of different heights and architecture, planted in an orderly manner form a multi stored and close cover vegetation with different biological cycles in agro - Horti / agro - Horti - silvi system.

In coconut and areca nut plantations other perennial or semi perennial and / or annual crops like tapioca, elephant foot yam, dioscrea, turmeric, ginger, colocasia, sweet potato, groundnut, vegetables and pulses, pineapple, bana, papaya can possible by grown. High density multi species systems are capable of generating high biomass; high income and can meet various needs of the farmer.

Silvi Pastoral System of Land Use: The fodder requirement for the growing cattle population in the country by 2000 - AD is expected to be around 700 mit against the present supply of 540 m.t. The gap will be of the order of 250 m.t. and to meet this demand the country will require 10 m ha of additional area. To achieve this every demanding increase of fodder the area under soil erosion problems, marginal and sub marginal lands waste lands village grazing areas need to planted with suitable forage species or in combination with economic perennial tree species preferably tree species yielding fodder.

In such land use systems, ideal species of woody perennials should be fast growing, hardy, with wide ecological aptitude, tight crown with multilayer branching and leaf orientation and of multiple use to the rural population. The forage component need to be very hard, easily colonising, palatable, nutritious and with strong establishment through roots or self sown seeds. For arid and semiarid areas species like Acacia tortillas. Albizzia amara, Hardiwickia binata, ledcaena leucocephala with Cenchrus ciliaris, Cenchrus setigerus, Dichanthium annufatum chresopogon fulvus sehima nervosum etc find greater adaptability. Legume species such as stylosanthes spp., have been found very versatile. On difficult sites Desmodium, Alcidcarpus and sesbania would promise as primary colonisers. Tourism and wild life system of land use.

Tourism as a means of expoiting and developing the potential of arid and semiarid areas needs careful consideration. The wildlife fauna of the arid region is unique resource and need attention for their conservation. It would help to develop more and more desert parks which attract the tourism and provide employment opportunities.

Bibliography

Andrews J.: *The Domesticated Capsicums*, University of Texas Press. 1995.

Arthur W. Gilbert, Mortier F. Barrus and Daniel Dean: *Growing and Breeding of Potatoes*, Asiatic Pub, Delhi, 2006.

Bander, J. : *Scheduling Irrigation with Evaporation Pans,* Montana State Univ., USA., 1984.

Bosworth, B.: *Capital Formation and Economic Policy.* Brookings Papers on Economic Activity, 1982

Brooks, D.: *Water: Local-Level Management*, Ottawa, International Development Research Centre, 2002.

Byers, Horace: *General Meteorology*, McGraw-Hill, New York, 1994.

Campbell, G.S. : *An Introduction to Environmental Biophysics,* New York, Springer-Verlag, 1977.

Chadha, K. L. and Pareek O. P.: *Advances in Horticulture: Fruit Crops,* Malhotra Publishing House, New Delhi, 1993.

Collinson, M.: *A History of Farming Systems Research*, CABI Publishing, UK, 2000.

David White: *The Physiology and Biochemistry of Prokaryotes*, Oxford University Press, Delhi, 2007.

Dudley, E.: *The Critical Villager: Beyond Community Participation*, London, Routledge, 1993.

Ferentinos L.: *Proceeding of the Sustainable Taro Culture for the Pacific Conference*, Honolulu, HITAHR, 1993.

Garret, J.R.: *The Atmospheric Boundary Layer*, Cambridge University Press, Cambridge, 1992.

Guthman J.: *Agrarian Dreams: The Parodox of Organic Farming in California*, University of California Press, Berkeley and London, 2004.

Hardy B.: *Biology and Agronomy of Forage Arachis*, Cali, International Centre for Tropical Agriculture, 1994.

Holton, J.R.: *An Introduction to Dynamic Meteorology,* Elsevier Inc., Burlington, Md, 2004.

Jacobson, Mark Z.: *Fundamentals of Atmospheric Modeling,* Cambridge University Press, New York, 2005.

Kovach, S.P. : *Injection of Fertilizers into Drip Irrigation systems for vegetables,* USA, 1984.

Lorenzen, S.: *The Phylogenetic Systematics of Freeliving Nematodes*, London, The Ray Society, 1994.

Marshall, F, R.: *Breeding Farm Animals*, Asiatic Pub, Delhi, 2006.

Merrill, Richard: *Radical Agriculture*, New York University Press, New York, 1976.

Nation, James L.: *Insect Physiology and Biochemistry,* CRC Press, 2001.

Oka H. I.: *Origin of Cultivated Rice*, Elsevier, Japan Scientific Societies Press, 1988.

Pemberton, R. W.: *Predictable Risk to Native Plants in Weed Biological Control,* Oecologia, 2000.

Pilgrim R.: *Post Harvest Handling of Minor Exotics*, St. George's, Grenada, 1996.

Qystein V. Sjaastad: *Physiology of Domestic Animals*, International Book Distributing Co., Delhi, 2005.

Rakesh Hooja *: Irrigation Drainage : International and National Perspectives*, Agrotech, 2000.

Reynolds, R.: *Guide to Weather*, Firefly Books Inc., Buffalo, New York, 2005.

Schilthuis, Willy: *Biodynamic Agriculture*: *Rudolf Steiner's Ideas in Practice*, Anthroposophic Press, Hudson, NY, 1994.

Tomi Petr *: Fisheries in Irrigation Systems of Arid Asia*, Daya, 2007.

Vanderplank, J.E.: *Plant Diseases: Epidemics and Control,* New York, NY, USA, Academic Press, 1963.

Watson, A.M.: *Agricultural Innovation in the Early Islamic World*, Cambridge University Press, Cambridge, 1983.

Index

❑❑❑